THE SPORTS CAR ENGINE
ITS TUNING AND MODIFICATION

THE SPORTS CAR ENGINE

ITS TUNING AND MODIFICATION

*

COLIN CAMPBELL
M.Sc., A.M.I.Mech.E., M.S.A.E.

1964

ROBERT BENTLEY INC.

18 PLEASANT STREET

CAMBRIDGE 39, MASSACHUSETTS

PRINTED IN GREAT BRITAIN BY
ROBERT CUNNINGHAM AND SONS LTD.
ALVA, SCOTLAND

Show me the mechanic who pops out his chest and announces 'I don't need any fancy gadgets to tune a car, all I need is to listen to the engine.' Show me this man and I will show you a hyper-egotistical DOPE.

Tom McCahill

Preface

I HAVE ATTEMPTED to write in one volume a textbook on the theory of automotive engine tuning, always with a bias towards the more highly developed type of engine as used in the modern sports car. Engine tuning is treated in its two aspects, basic tuning and super-tuning, i.e. the modification of a stock engine to increase the power output.

A supertuned engine operates with a tighter margin of safety than a stock engine and a loss of basic tune can soon lead to over-heating, even to catastrophic failure. With this in mind the major part of the book concentrates on the theory and practice of basic tuning. Chapters are devoted to the ignition system, to carburettors and to the methods used in 'trouble-shooting'. All of this is of value to the owner-mechanic, the garage man and the trainee racing mechanic. Supertuning is the subject of Chapters Nine to Thirteen. Here I attempt to take the magic out of supertuning. So many of the latest methods used in the British, German, Italian and American tune-shops are based on knowledge obtained in the engine labora-tories of universities, colleges and commercial research institutions throughout the world. Much of this knowledge is available to all of us, if we take the trouble to ferret it out. From time to time throughout the book the reader will see references to experimental work of this kind. The final chapter is on the mechanics of modifica-tion. By providing practical hints on the ways and means of engine modification this chapter should help to preserve the balance between theory and practice.

I am especially grateful to the editor of *Sporting Motorist* for permission to reproduce certain material used in Chapters Four, Five and Six, which appeared originally in that excellent journal as a series of articles on carburettor tuning.

COLIN CAMPBELL

Racine, Wisconsin, U.S.A.

Contents

List of Plates

CHAPTER ONE

General Principles

A PETROL ENGINE can start readily, run smoothly and give every appearance of being in good order, without necessarily being in good tune. The early motorist was happy enough if his engine started at all and kept running for the duration of his short journey. The modern motorist is more discerning and insists that his engine be maintained in an efficient state of tune, giving full power and a good fuel consumption. His day-to-day experience of his car's performance soon tells him when a tune-up is required. He remembers when a certain hill could be climbed in top gear. Since a drop to third gear is now necessary he is aware that his engine has lost its tune.

The choice of the word *tune* is very apposite, since there is a close analogy between the tuning-up of the instruments in an orchestra and the careful adjustments made to valve clearances, carburettors, plugs and points and ignition timing when tuning an engine to the peak of its efficiency. The term *tuning*, when applied to high performance engines, has a second meaning. If we raise the compression ratio, fit a carburettor to every cylinder, improve the design of the exhaust system, or fit a supercharger, we are still said to be *tuning* the engine. To prevent confusion it is better to call this *supertuning* or *modification*. In America they prefer more picturesque names, such as *souping* or *hopping-up*.

TUNING FOR PEAK EFFICIENCY AND POWER

Most of us are aware what engine tuning means in the average garage. New plugs and contacts are fitted. Valve clearances are checked—in a rather haphazard manner ('They never change much, anyway,' comments Bill Bodger, sagely). The carburettor idling screws are moved in and out and a squirt of whatever oil-can happens to be handy is put in the tops of all S.U. carburettors. This is the kind of thing thousands of so-called mechanics have been doing during the past fifty years. Some do it because they don't know any better; others because they couldn't care less.

If this is the simple painting-by-numbers approach that appeals to the reader, we are sorry he has been troubled. It is our firm belief that a real engine tuner must know the anatomy of engine behaviour as well as a good doctor knows the human body. In a primitive society we might be prepared to accept medical advice from men who had taken a two-week's course in First Aid. We cannot yet claim to be a civilised society, but we are at least a highly mechanised one. Our automobiles are an essential part of our modern lives and sports cars are a very pleasant adjunct to our leisure hours. It is only reasonable to insist that the men who tune these cars be specialists—real car doctors—not quacks.

Where does the power go?

Before we start to dig too deeply into the principles of engine tuning it will help in our basic training if we take a brief look at the thermodynamics of the petrol engine.

The energy in the fuel

When we burn a pound of fuel in a bomb calorimeter we burn it to completion and the whole of the available energy, i.e. the energy that can be released by combustion (not the atomic energy!), is released at this time and is calculated from the heat capacity of the system and the recorded rise in temperature. If we deduct the small amount of latent heat given up by the condensation of the steam—a product of the combustion of the hydrogen in the fuel—from the calorific value measured in the bomb, we are left with the nett calorific value. This is the total energy available in an internal combustion engine. For a typical petrol the value would be about 19,000 B.T.U. per lb. This then is the starting point for our thermo-dynamic considerations.

If we could design an engine to convert all the energy in the fuel into useful work and to transmit it to the driving wheels with no loss of power we could achieve a fuel consumption of about 150 m.p.g. cruising at 60 m.p.h. on a medium size sports car. In practice a figure of 35 m.p.g. at a steady 60 m.p.h. would be considered good.

The efficiency losses—step by step

In considering all the losses from the fuel supplied to the car-burettor to the power appearing at the rear wheels we can con-veniently group them under four headings:

1. cycle efficiency losses;
2. combustion losses;
3. pumping and friction losses; and
4. transmission losses.

Cycle efficiency losses

Many of the readers will be familiar with the concept, demon-strated about a hundred years ago by Joule, of a Mechanical Equivalent of Heat. Thus if we dissipate 778 ft. lbs. of work in friction we produce 1 B.T.U. of heat. By means of our heat engine, however, we are trying to convert heat energy into work and our efficiency of conversion is not 100 per cent. The percentage efficiency we can theoretically hope to achieve is dictated by the thermo-dynamic cycle on which we are operating our engine. It is custom-ary to consider the petrol engine as operating on what is called the Constant Volume or Otto Cycle. The Diesel engine is considered to operate closer to the Constant Pressure Cycle. In the first case,

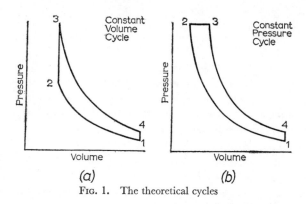

FIG. 1.   The theoretical cycles

the cycle begins at 1 in Fig. 1 (a) with a given mass of air which is compressed adiabatically (i.e. with no loss of heat to the surrounding walls) to the point 2. From this point the fuel is considered to be burned and the heat is released instantaneously to give a rise in pressure to point 3 with no change in volume. In practical terms, then, all the fuel would be burned while the engine remained at T.D.C. After the heat release has finished at point 3, the piston descends and the gas is expanded adiabatically to point 4.

The Constant Pressure Cycle is shown in Fig. 1 (*b*). This differs essentially in that the fuel is burned at such a rate as to maintain constant pressure in the cylinder while the gas is expanding between points 2 and 3. When all the heat release has occurred the piston descends and the gases expand adiabatically to the end of the piston stroke.

The petrol engine and the high-speed Diesel engine both operate on practical cycles that fall roughly between these two theoretical cycles. Moreover heat is lost to the cooling jackets during compression and expansion. Even so, if we were able to achieve the efficiency of the Otto Cycle, (which is slightly more efficient than the Diesel Cycle at the same compression ratio) we could still not convert all the thermal energy of the fuel into useful work. The reasons for this can be found in any standard textbook on thermodynamics. Here the reader can see evolved, if his mathematics are not as rusty as those of the writer, the following formula for the air standard efficiency of any engine operating on the true Otto Cycle:

$$E = 1 - \frac{1}{R^{k-1}} \qquad (1)$$

where E is the Otto Cycle efficiency

R is the compression ratio

k is the coefficient of adiabatic expansion for air, which is approximately equal to 1·4.

If we take a practical limit of 10:1 for the compression ratio, using modern premium pump fuel,

$$E = 60\%.$$

Thus we must inevitably lose 40 per cent of the energy of the fuel even if we succeed in achieving the maximum theoretical efficiency; that of the Otto Cycle.

Combustion losses

The initial rate of heat release at· or just before T.D.C. in a petrol engine must always be controlled by the degree of turbulence, by combustion chamber shape, by ignition timing, etc., to such a rate that the pressure rise between T.D.C. and the peak of the pressure curve does not exceed about 40 lb. per sq. in. per degree of crank rotation. If this value is exceeded by a substantial amount a harsh unpleasant mechanical 'roughness' will almost invariably be experienced. The modern phenomenon of 'rumble' is heard at pressure rise rates very close to 40 lb. per sq. in. per degree.

Recent work at the Joseph Lucas Laboratories has shown that this limit on pressure rise rate puts a serious limitation on the rate of heat release. Typical curves of cylinder pressure and heat

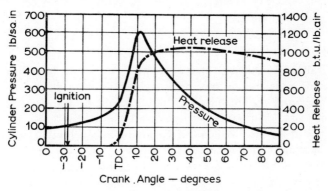

Fig. 2. Cylinder pressure and heat release during firing stroke

release rates are shown in Fig. 2. With ignition occurring 28 degrees before T.D.C. no measurable heat release occurs until 5 degrees before T.D.C. Ten degrees later, as shown by high-speed photographs of the inside of the combustion chamber, the flame had travelled all the way from the sparking plug to the far corners of the chamber. The whole of the combustible mixture in the cylinder was thus alight by 5 degrees after T.D.C. Even so, the amount of heat released at this time was only about one-third of the total achieved before the opening of the exhaust valve. At about 30 degrees after T.D.C. the rate of heat release reached a maximum and fell at a gradual rate until the end of the expansion period. Compared with the Otto Cycle then, a typical modern petrol engine burns its fuel at a relatively slow rate. Surprisingly, the loss of efficiency that can be attributed to this one factor of delayed burning is only about 5 per cent.

SCE B

Dissociation

At high temperatures some of the carbon dioxide molecules resulting from the complete combustion of fuel dissociate into carbon monoxide molecules. Other complex intermediate products, such as nitrogen oxides are always present during the combustion of petrol and air at high temperatures and pressures. The subject of dissociation and intermediate combustion products is too involved for a book on engine tuning. All the reader need know for the present purpose is that these reversible reactions waste energy and, for the particular engine operating conditions under discussion, can be considered as robbing us of about 14 per cent of the available heat energy.

Cooling losses

Ricardo has estimated that about one-tenth of the heat normally rejected to the cooling system through the walls of the combustion chamber and the cylinder and piston could be transformed into useful work. With the materials available to-day, however, it is difficult to see how we could risk running at much higher cylinder and head wall temperatures. The temperature of the piston crown and the ring belt too is far too critical in a high compression engine at full power for even a small reduction in the heat rejected to the cooling system.

Taking Ricardo's value of one-tenth of the total cooling system heat as recoverable, we can say that this represents about 5 per cent of our Otto Cycle efficiency figure of 60 per cent that we must consider lost in any practical engine.

The total combustion efficiency losses therefore add up to $5\% + 14\% + 5\% = 24\%$. Taking this from the original 60 per cent of the theoretical Otto Cycle we are left with an efficiency of 36 per cent. This is called the Indicated Efficiency, since it represents the amount of useful work that would be shown under the pressure–volume trace of an old-fashioned indicator diagram.

Pumping and friction losses

The indicator diagram only considers the useful work performed during the compression and expansion strokes. Some work is expended on a 4-stroke cycle engine in what is usually called 'pumping losses'. This is the amount of work involved in the induction process and in the exhaust removal. A good design of exhaust system should involve little power loss, but the average sports car engine at the chosen condition of 60 m.p.h. cruise will absorb about 5 per cent of the original fuel energy in pumping losses. Mechanical losses are relatively high at such a light load and we can consider about 7 per cent of our indicated efficiency of 36 per cent as lost in internal friction. About two-thirds of this friction loss can be attributed to piston friction.

Transmission losses

Our Brake Efficiency is thus $36\% - 5\% - 7\% = 24\%$. This represents the overall efficiency of the engine as measured in terms of energy transmitted at the flywheel. The efficiency of the transmission depends upon the gear in use at the time, being slightly lower for each drop in gear. With top gear engaged we can consider a good average figure as 92 per cent. Thus by the time useful work is performed at the contact patches of the driving wheels this percentage of the original fuel energy that appears as effective work is only $24\% \times 0.92 = 22\%$.

THE PENALTY OF POOR TUNE

Now let us consider what happens if the same car is in a poor state of tune. By 'poor tune' we do not mean that the engine is misfiring or has lost compression. This sad state of maintenance we would call 'a bad state of tune'. For the particular case let us consider that the air cleaner is dirty and has increased the pumping losses from 5 to 7 per cent. The ignition timing is also considered to have drifted about 5 degrees from the correct setting and the mixture strength is generally richer than normal from the choking effect of the dirty air filter. The indicated efficiency can be considered as reduced from the original 36 to 32 per cent. Since the pumping and friction losses have now increased from 12 to 14 per cent the new brake efficiency is $32\% - 14\% = 18\%$.

The transmission losses drop this to $18\% \times .92 = 16.5\%$. This is quite a drop from the 22 per cent of the engine in good tune.

In both cases we have considered the car as cruising at a steady speed of 60 m.p.h. When in poor tune the engine will require a larger throttle opening, more power and more fuel to maintain the same speed. For a car such as the Triumph TR3 the power required at the road wheels will be about 18 b.h.p. (see Fig. 3). The indicated horse powers required to be developed by the two cases to provide 18 b.h.p. at the driving wheels will be:

Good tune
$$\text{i.h.p.} = 18 \times \frac{36}{22}$$
$$= 29\cdot5$$

Poor tune
$$\text{i.h.p.} = 18 \times \frac{32}{16\cdot5}$$
$$= 35$$

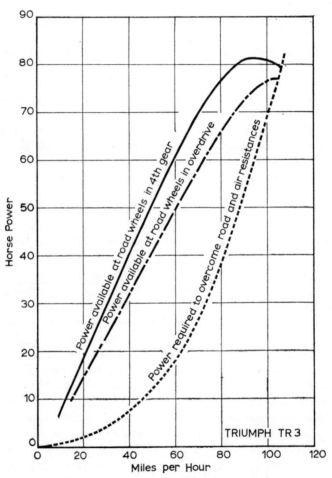

FIG. 3.   Power available and power required

The engine in poor tune, however, uses a disproportionate amount of fuel to provide its 35 i.h.p. The ignition timing is incorrect and the mixture strength supplied by the carburettors is enriched by the dirty filters. At the same indicated power then we can reasonably say that the poorly tuned engine will use about 10 per cent more fuel than the well tuned engine.

A Triumph TR3 in good tune will have a fuel consumption of 38 m.p.g. (imperial gallon). The fuel consumption of the car in poor tune will be in proportion to the extra indicated power demanded, with a penalty for the lower indicated efficiency of 32 per cent plus an additional penalty of 10 per cent for the mixture enrichment. The fuel consumption will thus increase approximately to:

$$38 \times \frac{29 \cdot 5}{35} \times \frac{32}{36} \times \frac{9}{10} = 25 \cdot 5 \text{ m.p.g.}$$

A similar calculation can be done to show how much loss of acceleration will result from the same loss of tune. Fig. 3 shows that at the engine speed corresponding to 60 m.p.h. in top gear a surplus power of $62 - 18 = 44$ b.h.p. is available when the engine is in good tune. The poorly tuned engine, however, not only requires a larger throttle opening to maintain the road speed of 60 m.p.h., but the power available at full-throttle will also be correspondingly reduced. It is not difficult to see from Fig. 3 that the cumulative effect of an increase in the indicated power required for steady speed and a decrease in the maximum available power can result in a loss of half the power available for acceleration.

Oddly enough, a moderate loss of power at high engine speed does not seriously effect the top speed of the car, since the speed of a car varies approximately as the cube root of the power. In the case considered the poor tune resulted in a loss in brake efficiency of 4 per cent, from 22 to 18 per cent. If this same power reduction, i.e. $4/22 = 18$ per cent is experienced at the peak of the power curve, the drop in top speed will only be 6 to 7 per cent. To the daily city commuter this will be of little consequence. To the competition driver it means failure.

Need we say more? Good tuning is as essential to the sports car as good health is to its driver.

SUPERTUNING

There was a time when motorists looked in awe at an engine that was different. There was something rather daring about planing a sixteenth of an inch from a cylinder head and the fitting of a supercharger could only be the prelude to 'dicing with death'. To-day the supertuning or modification of mass-produced engines has become a minor industry. In certain cases one can now arrange for a car to be modified before delivery and in such cases the makers still give a full guarantee to the modified car.

Before the war almost all manufacturers of cars objected to modification work, even when carried out by professionals. The manufacturer had usually spent a considerable sum of money in trying to extract more power from his standard product and had competed in all manner of international events with these modified cars. Despite this experience it was usually considered inadvisable to make this supertuning information available to the owners of the standard products. There was usually a haunting fear in the manufacturers' minds that widespread supertuning would inevitably lead to unreliability and the loss of a great reputation. At the time of writing it is nearly fifteen years since the M.G. Car Company issued their booklet *Special tuning for the M.G. Midget engine*. In the intervening years many thousands of enthusiasts have supertuned their Midgets, their MGAs and the new Midgets on the lines indicated in the appropriate M.G. tuning booklets. M.G.s tuned in this way have had successes all over the world in rallies, in speed hill climbs and in Club circuit racing. Occasionally a hard-driven supertuned M.G. has suffered engine failure, but the failures have been remarkably few and the reputation of the M.G. Car Company stands higher to-day than ever before.

Going it alone

With the spread of the tuning kit business to embrace almost every make and model on the market the young sports car enthusiast, having grown up in a world of tuning kits, tends to think of supertuning as the carrying out of a specific project with a comprehensive kit of parts and a full set of instructions supplied by the makers. We cannot dispute that this is one of the least expensive ways of increasing the performance of an engine, but a lot of the fun is taken out of supertuning when the thinking and the

planning has been done by someone else. Supertuning is largely
the application of basic physical laws, backed by a modicum of
engineering know-how. There is no reason why the more intelligent type of garage mechanic and the mechanically-minded
sports car enthusiast should not 'go it alone'. Many examples
from the past spring to mind. Raymond Mays, fresh down from
Cambridge, improved the lubrication system on the Brescia
Bugatti, thus lifting the reliable engine speed ceiling to new heights.
Even *Le Patron* had missed this modification! Later, Raymond
Mays took the 3 litre Vauxhall T.T. engine, an engine that was
already acknowledged as a phenomenal power-producer in the
1920s, and developed it, with the help of his two engineers, Villiers
and Jamieson, into his record-breaking supercharged Vauxhall-
Villiers.

Freddy Dixon, owner of a small Middlesbrough garage, worked
on the 1·5 and 2 litre Rileys in the mid-thirties and with the application of his motor-cycle experience plus an uncanny engineering
second-sight, he used the racing techniques of the fifties—tuned
induction and exhaust pipes—one carburettor per cylinder—'wild'
camshaft timings. Small wonder that the works entries often failed
to catch him!

A more recent example of the lone wolf is Dr Shephard in his
tuned 'stock' Austin A40, an expensive piece of private development work, but unbeatable in its class. We must also remember the
epic story of the Lotus organisation. Colin Chapman started as a
lone wolf, tuning and racing modified Austin Sevens.

The lone wolf often does things that the factory says are impossible. An example is the American treatment of the ports in the
cylinder head of the XPAG M.G. Midget. The *Tuning Handbook*
states that the stud boss in the centre of the siamesed port 'may be
ground away slightly—about $\frac{1}{16}''$ off each side (still maintaining its
streamline shape)—so that oblong ports are obtained, $1\frac{3}{16}''$ high,
$\frac{11}{16}''$ wide (minimum). Do not remove the boss completely or it will
affect mixture distribution.' When the American hot-rod boys saw
this boss they cut it right out and fitted a small streamlined nut to
hold down the head at this point. Perhaps they hadn't read the
Tuning Booklet! They enlarged the port area by about 20 per cent
at this point and even if they disturbed the mixtures no apparent
ill-effects were observed. They certainly made the TC Midget go.

The basic principles

We have stated that supertuning is largely the application of certain basic physical laws. What then are the basic rules of supertuning?

1. Make it breathe.
2. Make it efficient.
3. Make it reliable.

Making it breathe

The more air an engine breathes in a given time, the greater the horse power. This is a fundamental law of supertuning. The brute-force way to achieve this is to supercharge. This is so easy (up to the thermal and mechanical limits of the engine) that super-charged racing cars have nearly always been placed in a special class or raced against naturally aspirated engines of much larger capacity. There is more of a challenge in trying to get more air through a naturally aspirated engine. The downward movement of each piston on its intake stroke displaces a definite quantity of air and, if this air had no inertia and the connecting passages between the cylinders and the atmosphere offered no resistance to the flow of air the problem of filling the cylinders completely with air at atmospheric pressure would not exist.

Since the air has inertia, quite considerable at high speeds too, and the valves and ports and carburettor venturis all offer resistance to the flow of air it is the task of the tuner to make the inertia of the air work for him and to make a close study of all the obstructions in his induction and exhaust system that resist the free flow of the air into the engine and the free flow of the exhaust gas out. With turbulent gas flow, as exists in an induction system, the pressure drop across any restriction in the system varies as the square of the velocity. Thus if we can so modify the induction system that the total pressure drop from the atmospheric pressure at the air horn all the way to the inside of the cylinder is cut by one-third, the limiting gas velocity will increase to $\sqrt{\dfrac{3}{2}}$ times the original limiting velocity, an increase of about 22 per cent; the power output will increase in proportion. In this category of supertuning we can include all work to improve the flow characteristics of the valve

ports, the fitting of larger valves and the provision of special camshafts with increased lifts and modified valve timings to take
advantage of the increased inertia of the incoming charge at the
higher engine speeds induced by the improved valves and ports.
Also in this field we may include the fitting of larger carburettors,
larger venturis in the same carburettors or the provision of one
carburettor per cylinder. Here also we must include the profitable
field of exhaust and induction pipe 'tuning', whereby the overall
lengths of these two piping systems are 'tuned' to allow the high-
frequency pulsations that are present in both pipe systems to ram
additional charge into the cylinders over a desired range of operating speeds. The field is very large and will be discussed at length
later in the book.

Making it efficient

Every pound of fuel we burn in an internal combustion engine
has a certain potential for work. At light loads only a very small
percentage of this fuel performs useful work and at the most favourable conditions when the maximum brake efficiency is achieved

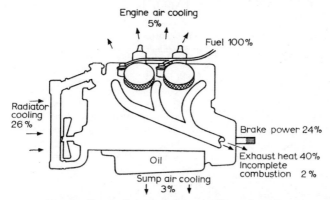

FIG. 4.   External energy balance at 60 m.p.h. cruise

less than 30 per cent of the fuel energy is converted into useful
work. The rest of the energy is wasted in losses to the cooling
system, to the air surrounding the engine, by miscellaneous frictional losses in bearings, pistons, valve gear, etc., and by heat
rejected at the exhaust system (see Fig. 4).

If we examine Equation (1) for the Otto Cycle efficiency we see that an increase in compression ratio will give us, in theory at least, an increase in efficiency. In practice this is found to be true, although for various reasons as discussed in Chapter Nine, practical engines do not gain as much from an increase in compression ratio as predicted by Equation (1). An increase in compression ratio from 9 to 1 to 10 to 1 would only give a gain in b.m.e.p. of about 3 per cent at the same engine speed. The speed at which the power curve peaks is usually increased slightly by the slight improvement in volumetric efficiency (at a higher compression ratio the ratio of hot residually gases retained in the cylinder after the exhaust stroke, to the weight of new charge is always less; this reduces charge heating and gives a higher volumetric efficiency). With the help of this increase in peak r.p.m. the peak power would be increased by about 4 or 5 per cent. Torque however, being proportional to b.m.e.p., would only increase by about 3 per cent. As we increase the compression ratio above 10 to 1, always supposing that knock-free fuels are available at these higher ratios and that the mechanical structure of the engine is robust enough to withstand

TABLE 1

Components	Power loss, as percentage of indicated power output	Action	Reduced loss
Fan	3%	Fit electric fan for low speed use only	0
Dynamo	1%	Disconnect for short events	0
Unbalanced rotating and reciprocating components	$\frac{1}{2}$%	Fully balance crankshaft, rods, flywheel	0
Bearings, piston rings, rockers, cams, water pump, timing chain, etc.	$7\frac{1}{2}$%	Carefully refit all parts, reduce oil drag	6%
Total	12%		6%

the higher internal pressures, the gains in efficiency become less and less for each step in unit compression ratio.

We can make our engine more efficient too by concentrating on the efficient working of the components that absorb power. A few examples are given in Table 1.

If all the above gains were achieved the mechanical efficiency of the engine would be increased from 82 per cent (taking the pumping losses at 6 per cent) to 88 per cent, giving an increase in brake horse power of $7\frac{1}{2}$ per cent.

Making it reliable

Reliability is a relative term. No car is absolutely reliable. The use of an engine of low specific power output did much to give the old Austin 12 and 20 a wonderful name for reliability and longevity. Not that good design is not essential if troubles with bearings and seals and gaskets and all manner of little things is to be avoided, but the deliberate restriction of power as a matter of policy reduces the stresses throughout the whole engine and cannot fail to give results. Thus we can state as a general rule of design: detune to increase reliability; supertune to decrease it. Only by excessive supertuning do we lose reliability completely.

The companies of the British Motor Corporation show a realistic approach in the Foreword which is common to all their *Tuning Booklets*. This is what they state:

> It must be clearly understood, however, that, whereas it is a simple matter to increase the power output of the engine, this increase in power must inevitably carry with it a tendency to reduce reliability. It is for this reason that the terms of the warranty of a new Healey Sprite (or other appropriate model) expressly exclude any supertuning of the kind described in this booklet, but this does not mean that tuning in this way will necessarily make the car hopelessly unreliable. In fact, it may be assumed that it will be at least as reliable as other cars of similar performance.

In assessing the cost of supertuning one should remember that on the credit side we are increasing the effectiveness of the car as a sports car. The acceleration is improved, the top speed is higher. The alternative to supertuning a sports car of modest engine size for road use is to sell it and buy a sports car with a bigger engine. (There are exceptions to this of course!) When the car is to be raced there is no alternative. One must supertune; all the other cars in the same class will be. Even in the Standard Production Class a modest amount of work is permitted.

When one buys a Birdcage Maserati one is paying for the manufacturer to supertune the car to its limit before delivery. This is an expensive solution which is only available to a few of us.

We can only consider each case on its own merits and the personal wishes of the owner of the car must exert a strong influence. Personal prejudice will often withstand a barrage of logical arguments. Why else would perfectly sane members of the Vintage Sports Car Club take well-preserved examples from the cream of European automobile history, strip them down, supertune them to give almost twice the original power output, then race them to near-destruction? Why indeed, unless it's for the sheer joy of motor racing.

CHAPTER TWO

Instruments

Good maintenance

Tuning and good maintenance are inseparable and maintenance of the chassis as well as the engine is a vital part of good tuning. This should be self-evident, but managers of racing teams have sometimes been known to neglect chassis maintenance. It is so easy for the enthusiastic amateur tuner to turn a blind eye on the rocker shaft brushes that are badly worn or the cams that have lost the case hardening from the nose. He can fiddle for hours with the carburettors and re-check the ignition timing day after day, but he will never recover the original tune of his engine until he restores the full lift of the cams by carrying out all the work of repair and replacement that we call maintenance.

Wherever possible, maintenance should be carried out to a plan. The maker's plan may not be perfect but it usually serves as a good start. If the maker's handbook states that valve clearances should be checked every 3000 miles, it is far better to carry out the operation when advised than leave the clearances unchecked until a valve burns out through total loss of clearance. Experience alone will tell if the recommended period of 3000 miles is much too frequent and whether it is safe to extend the period. Similarly, with all items of maintenance it is far better to check too frequently until experience is gained. One would not think of applying the planned maintenance technique used for a large fleet of vehicles to a single owner-maintained vehicle. A fleet owner must always balance the cost of too frequent inspections against the cost of breakdowns. In this way he plans the frequency of his maintenance inspections so that breakdowns are reduced to an economic low level. To try to inspect so frequently as to guarantee *no* break-downs would be extravagant.

It is therefore essential as part of the technique of keeping a car in excellent tune to evolve a system of maintenance. This system for its proper working requires a regular plan of inspections or tests and must inevitably involve periodic repairs and replacements when faults are revealed by these inspections. For example, our maintenance chart may call for re-packing the front wheel hubs with fresh grease every 10,000 miles. At the same time we check the wheel bearings for lift and rock. If play is considered excessive, the bearings are removed for inspection and, if necessary, are replaced by new ones.

TUNING INSTRUMENTS

Nobody can tune an engine accurately without instruments. Very few automobile engineers would argue with this to-day. The extent and the scope of the instrumentation considered necessary, however, is still a matter of individual choice. Even the experts differ widely here.

At this stage, before we launch into the broader aspects of testing and tuning, it will serve as an introduction to the concept of scientific tuning if we describe the instruments that are available to-day and discuss the interpretation of the readings given by them.

The dynamometer

The dynamometer or brake is such a large and expensive piece of equipment that one does not usually refer to it as a measuring instrument. Nevertheless its primary purpose is the measurement of torque and, in conjunction with an accurate tachometer, of horse power. With the addition of a fuel flowmeter, or fuel weigher, one can also measure specific fuel consumption, i.e. the weight of fuel consumed in producing one horse power at the flywheel for a period of one hour.

For research work on engines and for all manner of development work a dynamometer is almost essential. The effect of every small variable can be studied with accuracy. Fig. 5 serves to illustrate how a change of exhaust system can influence the power curve of an engine. A general increase in power could be readily assessed by road-testing, i.e. by acceleration tests or hill climbing tests against the stop-watch. One might even prove that Exhaust System A gave a higher maximum speed than System B, but without a dynamometer one would fail to learn the whole story of the two exhaust systems. Typical power curves, obtained by the writer on The Walker Manufacturing Company's 'Mid-West' dynamometer from a Pontiac Tempest engine, are shown in Fig. 6. Here we can see how the specific fuel consumption of the

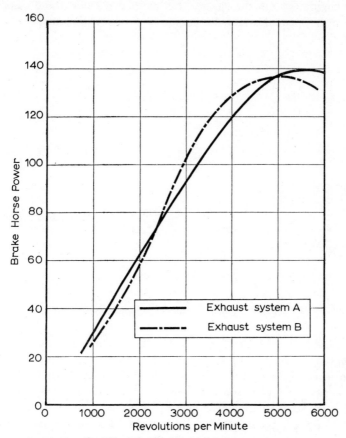

FIG. 5.   Effect of exhaust system on power output

engine varies, not only with engine speed, but with a gradual increase in throttle opening right up to full power.

Unfortunately, the use of a dynamometer is beyond the resources of the majority of engine tuners. Even if the initial cost of the dynamometer installation (at least £1000 for the minimum equipment to cover a range of 50-150 b.h.p.) is met by a commercial undertaking such as a garage or firm of experimental engineers,

very few motorists are prepared to pay £75-£100 for the cost of
removing an engine from the chassis, setting it up on a dynamometer stand, carrying out a quick series of power curves, and refitting
the engine back in the chassis. Let us then say that the dynamometer is essential when the engine is a new design and we are

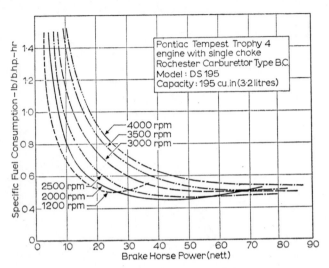

FIG. 6.   Power and fuel consumption curves

responsible for its development to the production stage. It is
essential when our business is the development and maintenance
of a team of Grand Prix cars or a top-flight team of racing sports
cars. For the typical sports car owner who wishes to supertune his
car to the stage where he can give a good showing in week-end club
racing there is really no need for a dynamometer. For those who
are interested in further reading on the subject, we can recommend
The Testing of High Speed Internal Combustion Engines by A. W. Judge.

Garage instruments

The instruments most commonly used in testing and tuning are
as follows:

tachometer (or rev-counter),	fuel pressure gauge,
vacuum gauge,	stop-watch,
compression tester,	voltmeter and ammeter.

Additional instruments of value to the garage specialising in engine tuning work are:

battery tester,
ignition timing lamp,
distributor analyser,
tach-dwell meter,
coil tester,
condenser tester,
sparking plug tester,
exhaust gas analyser,
stroboscope, and
cathode ray oscilloscope.

The first group are the basic tuning instruments, the essential tools of the serious amateur tuner. The second group are specialist instruments, and are on the whole more expensive. Although not essential, these specialist instruments simplify the diagnosis of the more obscure electrical and carburation faults. The stroboscope is most commonly used to study the running behaviour of the valve gear train.

The tachometer

For dynamometer work the accuracy of the tachometer is as important as the accuracy of the dynamometer itself. For garage tuning, no great accuracy is demanded; sensitivity being more important. An accuracy of 5 per cent is quite acceptable. The customary rev-counter on the sports car instrument panel sometimes falls outside the limits of $\pm 5\%$, but can still be of value if the sensitivity is good. This should be such that a drop in r.p.m. from 600 to 580 can be clearly seen. For the professional tuning establishment, Crypton Equipment Ltd in Great Britain and Sun Electric Corporation in America make a dual purpose instrument called a 'Tach-Dwell Meter'. This instrument serves both as an electrical tachometer and a meter for measuring the dwell angle of the contact breaker (the dwell angle being the angular movement in degrees from the closing of the points to the next opening).

No mechanical drive is required for this type of tachometer; an essential requirement for general garage use. The electrical impulses to operate the instrument are provided by attaching one lead from the instrument to the C.B. terminal and the other to earth.

SCE C

Further details are given later in the section dealing with specialist instruments.

The vacuum gauge

The vacuum gauge is as inseparable from the engine tuner as the stethoscope from the doctor. There is indeed a close parallel. Both instruments are simple in operation. Both require expert knowledge and experience before much value can be placed on the user's interpretation of the meaning of the readings or signals sent out. The vacuum gauge is simple enough to read, but mistakes can still be made in the interpretation of the readings, even if the operator is an experienced man. It is safer to regard the behaviour of the instrument in many cases as 'providing a clue'. With the support of other instruments or other tests the clue usually leads us to the right answer.

One could use a simple mercury manometer (a U-tube filled with mercury) to measure the induction vacuum, but this would be an inconvenient instrument in the garage. The popular vacuum gauge has a circular dial, calibrated in inches of mercury, and is operated by a Bourdon tube in the manner of a pressure gauge. It is connected to the induction manifold at any convenient point after the throttle-plate. To obtain a suitable air-tight connection at the manifold it is necessary to drill and tap the manifold on the engine side of the carburettor throttle. To be absolutely certain that no drilling can enter the engine one should remove the manifold before drilling and tapping, cleaning it carefully before refitting. A wise repair shop manager usually sees that a manifold adaptor is fitted during a top overhaul. If, however, it is not convenient to remove the manifold, the drill and tap should be liberally smeared with grease before use to make the chippings adhere to the tool. Before the final 'break-through' of the drill, the drill should be cleaned and greased again. There might appear an element of risk to some in this procedure, but the writer has not yet come across a case of engine damage being caused in this way. In any case, the reader has the remedy in his own hands, since he can remove the manifold if he so wishes.

Suitable manifold adapters can be purchased from Messrs Crypton or from the Redex company, or from the Sun Electric Corporation of Chicago. The Crypton adapter can be closed with a 4 B.A. screw or the adaptor can be removed after testing, and replaced by a 2 B.A. screw. Those cars fitted with vacuum operated windscreen wipers or windscreen washers already have a suitable connection on the manifold. Some engines are fitted with a manifold drain which makes a suitable vacuum connection. The connection between the adapter and the vacuum gauge can be conveniently made by thick-walled polythene tubing. Care is necessary in selecting the vacuum tapping to avoid places where the manifold is water-jacketed.

Vacuum tuning theory

When an engine is in perfect tune the induction depression or vacuum is at its highest reading for a given engine speed. Expressed another way, the higher we can make the vacuum gauge reading (without changing the engine r.p.m.) the more closely do we approach perfect tune. To explain this calls for a certain understanding of the physical factors producing the induction depression.

When an engine is idling, the throttle is almost closed. All the power developed by the engine is used in overcoming the friction losses in the bearings and in the piston rings and the pumping losses in induction, compression and exhausting. Only sufficient air is admitted at the throttle for this amount of power to be developed. If more air were admitted, more power would be developed and the idling speed would increase. The pumping losses and friction losses increase at higher engine speeds and with this wider throttle opening a new faster idling speed would be established giving equilibrium once more between internal and absorbed horse power. If an engine is not tuned to give maximum combustion efficiency —if for example, the ignition timing is too retarded or too advanced, or the mixture strength is too weak or too rich—a greater mass of air will have to be admitted past the throttle valve to develop the required idling power than would be the case with perfect tune. For a given engine, running at a constant idling speed, the more air we have to admit to the induction system to maintain this idling speed, the less will be the induction depression. Conversely, to maintain a high induction vacuum (low absolute pressure) the engine must operate on as little air (and fuel) as possible. To do this it must be in perfect tune.

Another conception that occasionally helps to elucidate this principle of vacuum gauge tuning is to regard the engine as a vacuum pump. A laboratory vacuum pump is usually driven by an electric motor. In this case, however, the motor is the petrol engine and the air to run this petrol engine comes from an air bleed into the vessel which we are trying to keep at a high vacuum (the induction manifold). Thus if we are to achieve a high vacuum in this vessel, the air bleed must be as small as possible. Since this is the air that runs the engine, the engine must burn this metered supply as efficiently as possible, i.e. it must be in perfect tune. Conversely, a high vacuum is an indication of a good state of tune.

One cannot use the vacuum gauge to compare the efficiencies of different designs of engine, only engines of identical type. The

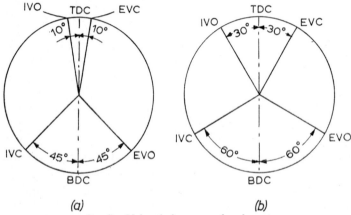

FIG. 7. Valve timings-normal and sports

complicated valve gear of a twin o.h.c. engine involves greater mechanical losses than the simple push-rod operated o.h.v. engine. Since the mechanical efficiency of the former is slightly lower at idling speeds the induction vacuum will also be lower. The use of large inlet valves for a high-speed engine calls for extra strong valve springs. To turn a camshaft against such springs involves greater wastage of power. Again, the use of large inlet valves involves a tendency to inadequate turbulence at low engine speeds. This reduces the efficiency of combustion at idling speeds. The valve timing has a certain influence on manifold depression. A typical saloon car valve timing would be as shown in Fig. 7 (a).

Where the designer is prepared to sacrifice low speed torque in order to extend the power curve to higher engine speeds, the exhaust valve opens earlier and the closing of the inlet valve is extended further up the compression stroke. Greater overlap of inlet valve opening and exhaust valve closing is also used (see Fig. 7 (*b*)). With such a camshaft pumping losses are increased at low speeds. The valve timing is unsuitable for low speed operation and causes a back-flow of gas into the induction tract before the

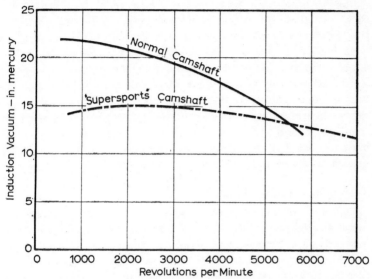

FIG. 8.   Influence of valve timing on induction vacuum

inlet valve closes. Unburnt mixture is lost into the exhaust system by virtue of the large valve overlap, since this is excessive for low speed operation. This type of camshaft, sometimes called a 'supersports' camshaft, will give its highest induction vacuum reading at a much higher speed than idling speed, usually between 1000 and 2000 r.p.m., and the maximum reading will not be as high as that given with a normal camshaft (see Fig. 8).

A Land-Rover, with an engine designed to give good low speed torque, will give a vacuum gauge reading of as much as 23 inches at 500 r.p.m. A more normal sports car engine such as the Austin-Healey will give 22 inches at 600 r.p.m. The Porsche 1600 Super, with large-overlap camshaft and relatively large inlet valve gives a maximum vacuum reading of about 15 inches, this maximum occurring at about 1500 r.p.m.

Once it is accepted that the highest attainable vacuum reading gives maximum efficiency the tuning of the carburettors or the timing of the ignition becomes simplicity itself. In the case of the ignition timing, the distributor clamp bolt is slackened back while the engine is running at a fast idling speed of about 800 r.p.m. The distributor body is slowly turned in the 'retard' direction until a definite drop in the vacuum reading is observed. The distributor is then rotated slowly in the opposite direction until the maximum reading is passed and the reading is seen to fall again by being too advanced. Now that the 'feel' of the thing has been obtained, it is a simple matter to set the distributor at the position giving the maximum reading and lock the clamp bolt. Similarly the carburettor slow-running mixture screw (or the jet adjusting nut on the S.U. carburettor) can be set to the correct mixture strength by screwing in and out until the setting is found which gives the optimum vacuum reading. The difference between tuning for maximum power and tuning for economy will be discussed later.

Testing and trouble tracing

Before tuning an engine one should test it. In this way, faults are sometimes discovered before the final tuning adjustments are made. Sometimes we are aware of a fault in the running of an engine before it is tested. Perhaps the engine is known to misfire at high speeds, or it spits back through the carburettor, or—more vaguely—the owner complains of a falling off in power. It is in this important work of trouble tracing that the vacuum gauge can sometimes be very useful. This aspect of its use is covered in Chapter Eight.

The compression pressure gauge

This instrument is used solely for testing. Where we once felt the resistance of each compression on the starting handle, we now use a compression gauge. It is much more accurate and, when the correct technique is used, is more revealing.

The compression tester is a simple but accurate Bourdon tube type of pressure gauge. The range is usually 0-200 lb. per sq. in., but an upper reading of 250 lb. per sq. in. is advisable for work on sports/racing engines. A non-return valve and pressure release device is incorporated, This permits the maximum reading obtained on each cylinder to be recorded before the pressure is released ready for the next test. The Redex Company of Chiswick High Street, London, make a convenient model with a flexible connection between the gauge and the cylinder adapter. In this model the cylinder adapter is threaded to fit the sparking plug hole. The Crypton compression tester is quicker in use, the gauge being fitted with a robust rigid pipe attachment on the end of which is a conical rubber bung. This bung is pressed tightly into the plug socket while the test is carried out. Where one or more plugs are badly situated for access of a rigidly mounted gauge the 'Redex' gauge, with its flexible connection is invaluable. The manner in which 'dry' and 'wet' compression tests are carried out and the interpretation of the results is discussed in Chapter Seven.

The fuel pressure gauge

To check the performance of a petrol pump it is necessary to measure the pressure given by the pump and the rate of delivery. Pump pressure is conveniently measured by means of a 0-5 lb. per sq. in. gauge. For an electric pump, such as the British S.U. pump, there is no need to run the engine to check the pump. A connection is made between the pump outlet and the pressure gauge, the pump is switched on and the maximum reading recorded. In the case of a mechanically driven pump such as the A.C., the engine can run on the fuel in the carburettor bowl while the pressure is measured. The pump should hold a steady pressure for at least ten seconds after the engine is switched off. An S.U. pump Type L in good condition should give a pressure of at least $\frac{3}{4}$ lb. per sq. in. No pulsing should occur after the pressure has built up to its maximum. An S.U. High Pressure pump in good condition should give a pressure of at least 1·0 lb. per sq. in. and an A.C. mechanical pump is considered acceptable with a delivery pressure of $1\frac{1}{2}$ lb. per sq. in.

Fuel pump delivery rate is checked by attaching an open ended pipe to the pump outlet (the pipe used for the attachment of the pressure gauge can be used if its bore is no less than the feed pipe to the carburettors). By means of a stop-watch the time to pump $\frac{1}{4}$ pint of fuel is measured, the fuel being caught in a suitably graduated vessel. In the case of the engine-driven A.C. pump the sparking plugs should be removed and the engine turned over on the starter. A Low Pressure S.U. pump should deliver this quantity in no more than 30 seconds, the H.P. type in no more than 22 seconds. An acceptable figure for the medium-size A.C. pump is 40 seconds.

The stop-watch

This is essential when we come to check performance on the road. As already stated, it is required when we check the output of the fuel pump. Robust construction is more important than accuracy.

The voltmeter and ammeter

For testing the ignition system and the dynamo a good voltmeter is required. A useful general purpose instrument to cover both 6 and 12 volt systems would be one with a range of 0-20 volts. This voltmeter should be of greater accuracy than the usual low grade instrument fitted to an automobile instrument panel. A high grade moving-coil instrument giving an accuracy of $\pm 0 \cdot 5$ volts is required, if we are to gain anything of value from its use. A sturdy ammeter is a useful garage instrument. A general purpose instrument would have two ranges, 0-10 and 0-100 amps.

To measure the cleanliness and good fit of the contact breaker points, it is advisable to measure the voltage drop across them. With well-ground points this should not exceed $0 \cdot 1$ volts. To measure this with any accuracy a special low-reading voltmeter is required.

SPECIALIST INSTRUMENTS

The second group of instruments described in this chapter are not those one would expect to find in the garage of an enthusiastic amateur. Several of them would not even be found in the usual public garage or repair shop. For the garage proprietor who wishes to carry out serious tuning work on customers' cars the following instruments are all well-tried and valuable aids to help him in the quick diagnosis of engine faults. None is absolutely necessary; but all are useful.

The battery tester

The popular type of battery tester in public garages is the heavy discharge cell tester. Within its limitations it tests the starting ability of a battery. Each cell is shorted out in turn through a low resistance path and the voltage during the discharge period noted. If the voltage registered by the cells falls below 1·5 volts the battery is not considered to be in a good enough condition to fulfil its duty as a starter battery. If a cell is observed to drop 0·2 volts or more below the readings of the other cells, this cell is condemned as defective.

As a means of boosting the sale of batteries the heavy discharge tester is an excellent instrument. To the conscientious garage man, however, condemnation by this instrument should not be taken as final. The resistance incorporated in the instrument is of fixed value. Thus if this resistance is chosen to represent the discharge taken from a large 15 plate battery when starting a 4 litre engine, it will take an excessive discharge from a small 7 plate battery, the size fitted to small-engined cars. The design of the heavy discharge tester can only aim at a compromise and the result is a test which is too light for a large capacity battery and too severe for a small one. There is also a second way in which this tester can be misleading. If the battery is in a partially discharged condition, as might occur if a customer has had difficulty in starting his badly-tuned car on a cold morning, the heavy discharge tester will show a large voltage drop. If a hydrometer reading below 1·200 is given by a battery, the heavy discharge tester should not be used until the battery has been re-charged. Some modern rapid battery chargers use an improved design of discharge tester with a variable resistance. Moreover the discharge tester is only used *after* the battery has been on charge.

For testing the amount of charge in a battery the hydrometer is still the most dependable instrument. A fully charged battery should give a reading of 1·280 and a fully discharged battery 1·130. A modern development from the simple hydrometer is the Crypton 'Hydro-lek' meter. By means of a Wheatstone bridge circuit this instrument can read each cell voltage on open circuit to an accuracy of ±0·001 volts. This is as accurate an indication of the state of charge as can be given by a hydrometer and can be carried out in a fraction of the time. With this instrument one can decide whether a battery needs to be put on charge. After charging at the recommended rate and for the recommended period the modern discharge tester, with variable resistance, can be relied upon to indicate whether the battery is sound or not. It is obvious that a well-tuned engine is of little use if it cannot be started. It is not always realised though how much we depend on the battery for good tune.

The modern high speed engine demands a good battery for the satisfactory working of the ignition system. If the voltage measured across the coil falls below 10 volts on a 12 volt system the spark intensity is liable to prove inadequate at engine speeds above 5000 r.p.m. The battery then is our starting point in checking the tune of an engine and it is at this first stage that the experienced tuner often finds the cause of that mysterious misfire at high speed. The untrained man changes the plugs, coil and condenser, strips the petrol pump and cleans all the carburettor jets before he thinks about the battery.

Systematic checks, carried out in the right order, are the essence of any sound tuning and testing technique. The writer has witnessed so many failures caused by lack of system that he feels he cannot repeat too many times the maxim: *Test before tuning and omit no tests.*

The ignition timing light

The vacuum gauge offers the quickest method of setting the ignition timing. Unfortunately, it is not the most accurate. Two methods are open to us for making more direct measurements. The old method of the 'pea-lamp' was accurate to about ± 2 degrees. A bulb of the same voltage as the battery is connected between the terminal on the base of the distributor and earth. When the contacts are closed the bulb will be out, since the current will flow through the path of least resistance, across the contacts. By turning the engine very slowly, beginning on the compression stroke of No. 1 cylinder, and stopping immediately the bulb lights up, examination of the timing marks on the flywheel (or on the crankshaft pulley) will show if the ignition is correctly timed. This static method is now largely superseded by the use of the more accurate stroboscopic light, such as the 'Synchrolite', made by Runbaken Ltd of Manchester. In this method a neon lamp is connected in series with No. 1 sparking plug lead. The flywheel inspection plate is removed (where such exists) and a spot of white paint dabbed where the ignition timing mark is stamped on the flywheel. In certain cases only T.D.C. for No. 1 cylinder is marked. In such cases it is necessary to mark out on the periphery of the flywheel the distance represented by the number of degrees of advance, as stated in the manufacturer's manual. If the number of teeth on the starter ring is known, the angular spacing between these can be used as a measure. For example if the number of teeth is 120 the angular spacing is 360/120 or 3 degrees. If the recommended static timing is 5 degrees before T.D.C. the white paint spot should be placed $1\frac{1}{2}$ teeth before T.D.C.

The method of checking the ignition timing with this instrument is very simple as soon as the timing mark is established. With the engine idling at about 500 r.p.m. the timing lamp is held so that the high frequency flashes shine on the face of the flywheel. Since these flashes occur every two revolutions made by the flywheel (on a 4-stroke engine) the flywheel appears to be stationary under the light of the lamp and the relative positions of the flywheel marking and the stationary pointer can be clearly seen. As the engine speed is increased we can watch the effect of the vacuum advance unit (if fitted) beginning to work. As the throttle is opened the small drilling connected to the vacuum pipe becomes uncovered and the full engine depression is transmitted to the vacuum unit on the distributor. This sudden ignition advance can be observed by the stroboscopic timer and is a quick check on the working of the vacuum unit. At higher speeds it is not easy to separate the effects of the vacuum advance unit and the centrifugal advance mechanism. This problem does not arise, of course, when a vacuum unit is not fitted. A simple lamp of this type is therefore useful as a rough-and-ready means of checking that the centrifugal advance is occurring. For an accurate check of the distributor advance characteristics one must use a distributor analyser or the latest stroboscopic method marketed by Crypton Equipment Ltd which is called the 'Motormaster'.

The distributor analyser

This is a specialist instrument and is not in common use in Great Britain. It probably owes its popularity in America to the prevalence of the V-8 engine, since the synchronisation of the dual contact breaker used on all American V-8s is an extremely difficult operation without the assistance of this special equipment. In use the distributor is taken from the car and mounted on the analyser where it is driven by a variable-speed drive from an electric motor. Any speed can be selected at will, from idle to 6000 r.p.m. equivalent engine speed. A stroboscopic ignition indicator and a vacuum pump are built into the unit. With the help of these the whole range of the speed advance curve and the vacuum advance curve can be studied and compared with the maker's specification. The following additional measurements can be made: distributor resistance, contact spring tension, dwell angle, dwell variation. Tests can be made to check for cam bounce, contact alignment and shaft and bush wear. These analysers are rather costly and their purchase can only be recommended to establishments with a large potential tuning business. In America the Sun Distributor Tester, Model DT-600 can be recommended. Crypton Equipment Ltd market a similar analyser in Great Britain. Their latest model, the B75, embodies many years of experience in the design of such units.

The tach-dwell meter

This is an electronic instrument that conveniently serves a dual purpose. It counts the electrical impulses from the opening of the ignition contacts and gives a direct reading in engine r.p.m. This is the principle of the popular electronic tachometers that are now fitted to many modern sports cars. The Tach-Dwell meter can also measure the percentage of total time devoted to contact dwell and give a reading in terms of dwell angle. Variation in dwell angle over the whole speed range can also be observed as a check for shaft and bush wear.

The coil tester

Many of the tests carried out by this instrument are within the scope of standard electrical test meters, but the professional engine tuner will still find that the speed and convenience of this specialist instrument soon pays for its cost. With it he can check coil stall current against a calibrated load. He can check the windings for continuity. He can check the output of the coil 'cold' and 'hot' and measure the resistance of the secondary winding. The writer has found the Crypton coil tester to be of great value in his work.

In America, the Sun Model CCT-222 is a good general purpose coil tester and can be used to check distributor caps, rotors and plug leads for insulation breakdown.

The condenser tester

The capacity of the condenser, measured in microfarads, is indicated by this instrument. Series resistance and leakage can also be measured.

Coil and condenser testers are sometimes supplied as components of a full test-panel, such as the Crypton Motor Analyser. The quick tune-up kits such as the Crypton 'Motormaster' and the Sun TUT-200 and STUT-300 models do not include these instruments, but they can be purchased as separate instruments from either firm.

The plug tester

All the well-known sparking plug manufacturers sell plug testers, usually as part of their plug cleaning rig. The plug is inserted into a small compression chamber, which can be pressurised gradually by turning a valve on the panel (garage compressed air supplies the pressure). A high-tension spark system supplies a continuous spark to the plug. The pressure at which the plug fails to spark is a measure of the condition of the plug. For demonstration purposes to the customer, a second plug socket is usually provided to take a new plug. Many car owners suspect the plug tester as a form of confidence trick to boost the sale of new plugs. In the writer's experience this suspicion is not justified by the facts.

The exhaust gas analyser

The traditional portable laboratory apparatus for the determination of the percentages of carbon dioxide (CO_2), carbon monoxide (CO) and oxygen (O_2) present in the exhaust gases of an internal combustion engine is the Orsat apparatus. CO_2 is measured by absorption in potassium hydroxide, CO by absorption in cuprous chloride and O_2 by absorption in potassium pyrogallate. This apparatus is still one of the most reliable of test-bed methods of exhaust gas analysis. Unfortunately it is too delicate and too cumbersome and far too slow for garage use. It also requires standards of cleanliness that are difficult to achieve in a garage.

The manner in which the CO_2, CO, and O_2 content of the

exhaust gas from a single cylinder engine varies over a wide range
of mixture strengths is shown in Fig. 9. Thus a large excess of CO
in the exhaust gas indicates a rich mixture and a large excess of O_2
indicates a lean mixture. On a multi-cylinder engine, however, a
large excess of O_2 can also indicate that one cylinder is misfiring
(watch out for that trap!). When the exhaust gases from a multi-
cylinder engine, with all plugs firing well, are sampled from a common pipe there are still variations present in the exhaust gas compositions from different cylinders. Distribution inequalities usually
make some cylinders richer, others weaker, than the mean. A

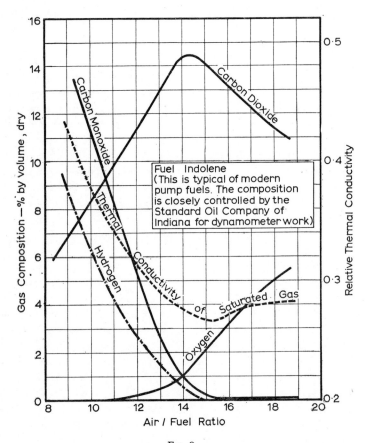

Fig. 9

multi-cylinder analysis that shows 4% of CO, 4% of O_2 and 12% of CO_2 would be an indication of poor distribution.

Despite this variation from cylinder to cylinder the overall air/fuel ratio can still be estimated with fair accuracy from the overall analysis of the mixed exhaust gases. The nomograph given in Fig. 10 is due to L. S. Leonard of the Ford Motor Company of America (S.A.E. paper No. 379A, *Fuel Distribution by Exhaust Gas Analysis*, read at the 1961 Summer Meeting). The writer has found

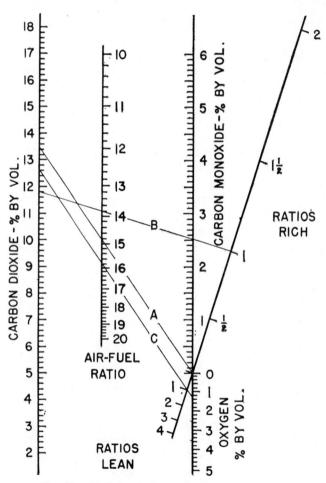

FIG. 10. Air-fuel ratio from exhaust gas composition

this nomograph very helpful in interpreting exhaust gas analyses
taken during his dynamometer work.

So far as the writer is aware all garage type exhaust gas analysers
operate on the same principle. They measure the differences in the
thermal conductivity of the gas and interpret this to indicate the
relative amounts of hydrogen and carbon dioxide present. Nitrogen,
oxygen and carbon monoxide have thermal conductivities close
to that of air. Hydrogen conducts heat nearly seven times as fast
as air; carbon dioxide about half as fast. It is thus the amount of
hydrogen present that is the major factor and this makes all these
instruments more dependable for mixtures richer than chemically
correct. Lean mixture indication is not reliable.

Fig. 11 shows how a Wheatstone bridge is used in a typical gas
analyser to compare the thermal conductivity of the sample gas

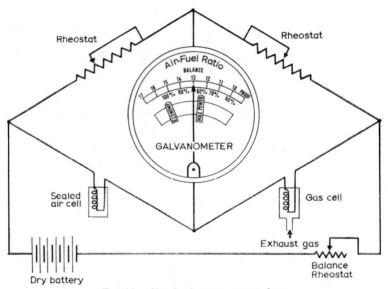

FIG. 11.   Circuit of crypton gas analyser

with the sealed comparator air cell. The temperature of the heated
filament in the gas cell will vary with changes in the composition
of the exhaust gas, being lowest when the hydrogen content is high
and highest when the carbon dioxide content reaches its peak at
an air fuel ratio of about 14·5 to 1. Fig. 9 shows that the thermal

conductivity of the saturated exhaust gas is at its lowest when the CO_2 content is close to its peak. It is also apparent from Fig. 9 that any exhaust gas analyser that interprets air/fuel ratios in terms of the thermal conductivity of the gas must give meaningless answers when the mixture strength is weaker than 15 to 1, since hydrogen is no longer present and carbon dioxide has passed its peak.

Fortunately for us, the sports car tuner is seldom interested in weak mixtures and for this type of work the writer can thoroughly recommend this type of instrument for garage use. One word of warning: these analysers should always be used with an efficient water-trap to remove all droplets of condensate from the exhaust gas sample that is fed to the analyser. Not all condensers supplied by the makers of these instruments are adequate. The large finned condenser supplied with the Sun analyser is a fine design and is well worth copying. The presence of water droplets in the gas stream is indicated by sudden kicks on the galvanometer needle towards the rich end of the scale.

For road-testing of carburettor modifications the writer has found the Crypton analyser to be a wonderful time-saver. A flexible hose can be clipped to the tail-pipe and brought in through the front window to the instrument. For safety it is advisable to take an assistant to read the instrument, taking care at the same time that the car interior is well ventilated, since the instrument exhausts the test gas into the car.

The stroboscope

Not to be confused with the stroboscopic ignition timing light, this general purpose type of stroboscope is operated by mains electricity to provide a powerful intermittent flash from a high candle-power lamp. The frequency of the flashes can be varied at will by means of a manual control. Thus if we point the light at the overhead valve gear when the engine is running at about 2000 r.p.m. and adjust the frequency gradually to about 1000 flashes per minute until the valve gear movement appears to be almost stationary, the opening and closing of the valves, any irregularities in the behaviour of the springs, any tendency to valve bounce, any of these failings can be studied in slow motion. This is only an example. The instrument has a wide application to

SCE D

moving machinery and can be used to study the motion of such things as the contact breakers, the fan pulleys, the propeller shaft and universals, in fact anything that moves to a regular pattern.

The cathode ray oscilloscope

The C.R.O. or 'scope was first applied to internal combustion engines as a combustion research instrument. A special cylinder head with an additional socket to accommodate the pressure

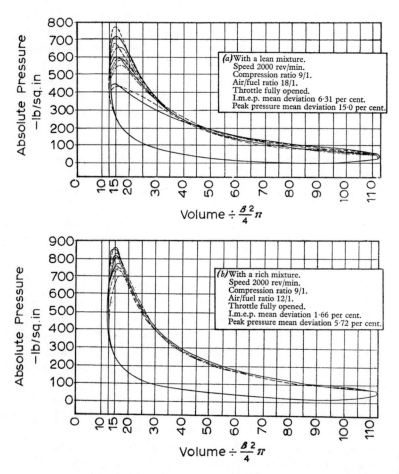

(a) With a lean mixture.
Speed 2000 rev/min.
Compression ratio 9/1.
Air/fuel ratio 18/1.
Throttle fully opened.
I.m.e.p. mean deviation 6·31 per cent.
Peak pressure mean deviation 15·0 per cent.

(b) With a rich mixture.
Speed 2000 rev/min.
Compression ratio 9/1.
Air/fuel ratio 12/1.
Throttle fully opened.
I.m.e.p. mean deviation 1·66 per cent.
Peak pressure mean deviation 5·72 per cent.

FIG. 12. P-V diagrams for ten consecutive cycles

pick-up was originally required. To-day we have special combination sparking plugs and pick-ups that allow us to obtain cylinder
pressure indicator diagrams from standard engines with only one
sparking plug socket per cylinder. Fig. 12 shows typical pressure/
volume traces obtained from one cylinder of a multi-cylinder research engine.

The improved reliability and versatility of the modern 'scope
made it possible to introduce it in recent years as a garage ignition
system tuning and trouble-shooting instrument. These ignition
'scopes are made now by many firms on both sides of the Atlantic.
The Heath Company of Benton Harbor, Michigan, are even
marketing an easy-to-build Heathkit 'scope at a very reasonable
price.

When clipped into the ignition system by detachable clips the
ignition pattern from any sparking plug can be studied separately.

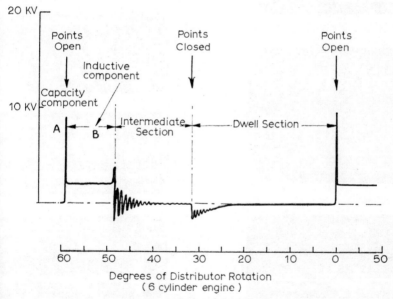

Fig. 13.   Ignition scope diagram

Alternatively, all the plug patterns can be superimposed. Fig. 13
shows a typical trace with a good plug and a good ignition system.
The trace can be considered in four sections. The firing section A

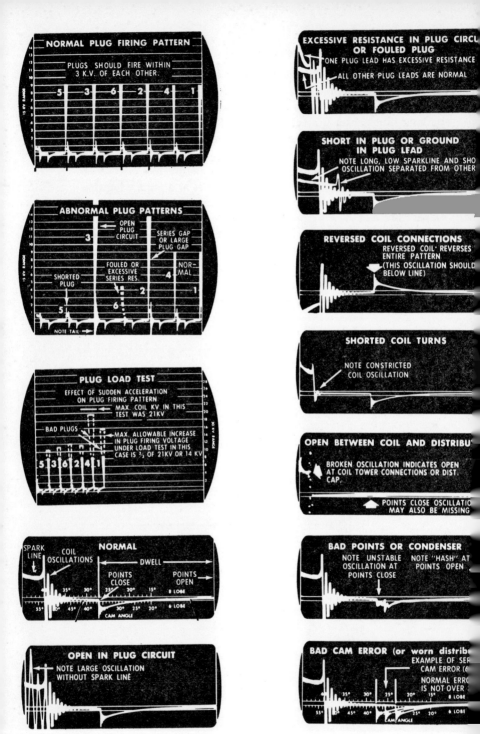

Fig. 14. Oscilloscope traces: a few ignition faults

is the capacity component of the spark, a high voltage transient of extremely short duration. The firing section B is the inductive component, the visible part of the spark. This is a lower voltage discharge that continues for a measurable period after the spark gap has become ionised by section A. (A more complete discussion of the spark phenomena is given in the next chapter.) The third section of the trace is usually called the intermediate section. After the spark has become extinguished there remains in the secondary windings a certain electrical energy which is dissipated as an oscillatory interchange between primary and secondary windings. Except at very high speeds, this oscillation has died out before the fourth section, the dwell section, starts. The dwell section starts with the closing of the contacts. At this time current begins to build up again in the primary windings. This creates a momentary oscillation in the secondary circuit.

Fig. 14 illustrates some of the faults that can be revealed by a study of the ignition traces from a full set of plugs. Other tests can be made to study the behaviour pattern of the condenser, the coil and the distributor. No other instrument can analyse the behaviour of the ignition system as accurately as the ignition 'scope, but the availability of the instrument is only half the story. It is the ability of the operator to make an intelligent evaluation of the traces he sees that decides the success or failure of this new technique.

CHAPTER THREE

The Ignition System

COIL IGNITION is almost invariably used to-day on modern sports cars. Magnetos are occasionally fitted as special equipment when tuning an engine to give exceptionally high rotational speeds, but in general the high voltage coil and the modern distributor is adequate for the engine speeds achieved by to-day's sports car engines. The design of modern magnetos will be discussed at the end of the chapter.

In recent years a minor revolution has been occurring in the field of automotive ignition. One cannot help wondering why this spate of inventive genius is happening all at once, after such a long period of relative stagnation. First transistorised ignition, then electronic ignition, and, overlapping the two in time, piezo-electric ignition. At the time of writing these are all in the experimental stage and sufficient experience has not yet been accumulated to permit the writer to attempt to give advice on the maintenance and tuning of such equipment. In Chapter Eleven a general survey is made of special tuning techniques and special equipment. A description of these new types of ignition system will be found there.

Ignition requirements

The ignition equipment on a modern petrol engine is more susceptible to maladjustment than any other part of the engine. It is therefore the most common cause of an engine losing its tune. The reasons for this will become more apparent when we examine the exacting demands made upon the equipment and study its inherent limitations.

To ignite the compressed mixture of petrol and air in the cylinders requires a spark to cross the plug points at the appropriate times. To produce the spark in a highly compressed gas the voltage must be very high. With a very small gap the voltage required would be much lower, but a gap of at least 0·015 inches is necessary to provide sufficient heat energy in the discharge to ignite the wet mixtures that exist in the cylinder when starting from cold. The gap is usually set in the range 0·025-0·035 inches. Sometimes, when the carburettor is tuned to give weak mixtures during cruising conditions, the gaps are set as wide as 0·040 inches. With such a gap the secondary circuit is required to provide a voltage of 10,000 to 15,000 volts to trigger the spark. This high voltage must be produced from a low voltage source, a battery giving a nominal voltage of 12. This is a step-up ratio of about 1000 to 1.

The primary circuit

The primary circuit consists of the battery, ignition switch, coil, primary winding, contact breaker and condenser, as shown in Fig. 15. With the ignition switched on, the rotation of the contact breaker cam causes the points to close and a current to flow from the battery, through the primary windings, contact breaker points and back to the battery by means of earth (the metal parts of the car). This produces a magnetic field around the soft iron core in the centre of the coil. When the engine is running at speed the available time for the build-up of the current and the subsequent build-up of the magnetic field is extremely small. Moreover the self-induction effect of the coil opposes any change in current strength and therefore resists any rapid build-up of current. On a 6-cylinder engine at 4000 r.p.m. the available time is approximately three thousandths of a second. This is insufficient time for the current or the magnetic field to reach their maximum steady values and, as will be seen later, this time factor can be a critical limitation to the satisfactory working of the system at high engine speeds. When the points break, the self-inductance of the coil again opposes the change and without the addition of a condenser to the circuit the current would arc across the points. The behaviour of the condenser may be likened to the buffers in a railway station which store up the energy imparted to them by the train and return it after the train has been brought to rest. A condenser of adequate size will store the electrical energy in its plates after the points break and after reaching its maximum charge density will discharge back to the primary windings.

It is often stated that the condenser is fitted to prevent burning of the points. The real purpose of the condenser is to provide a reservoir for the free flow of the charge from the primary windings. Without this alternative path, discharge would take place across the remaining high resistance path, i.e. by arcing across the points. This would not give us a rapid enough change in flux density in the magnetic field around the soft iron core of the coil. With a well-designed and well-tuned ignition system collapse of the magnetic field can generate a transient voltage of 250 to 300 volts in the primary windings. If this primary voltage is reduced for any reason, then the voltage in the secondary circuit is also reduced.

The secondary circuit

The secondary circuit consists of the coil secondary windings (containing about a hundred times as many turns as the primary windings), the distributor and the sparking plugs (see Fig. 15). In

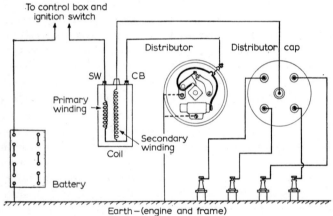

FIG. 15.   Ignition system circuit diagram

the modern ignition system the name 'distributor' has come to be used for the whole mechanism which houses both the low tension contact breaker and the high tension distributing device. This latter section comprises the rotor, which is driven by the same shaft as the contact breaker cam, and the distributor cap, which distributes the high voltage current to the sparking plugs in the correct sequence.

The spark occurs at the plug points when the voltage build-up in the secondary windings has reached the value needed to jump the gap. This may be anything from 5000 to 20,000 volts, depending upon the size of the gap and the pressure in the cylinder at the moment of discharge. As stated earlier the spark is not instantaneous, although the duration is not much more than a thousandth of a second, or two or three degrees of crankshaft rotation. During this time the discharge of the energy stored in the condenser initiates an oscillating current in both primary and secondary circuits which persists until the spark energy is dissipated. As soon as the contact points are closed again the current starts to build up again in the primary circuit and the whole process is repeated for the next plug to be fired. For a 6-cylinder engine running at 5000 r.p.m. the time available for the whole sequence of events is about five thousandths of a second (see Fig. 13).

Secondary voltage variations

The voltage build-up in the secondary circuit is lower at high speeds than at low. The modern coil is specially designed to reduce the time required for the magnetic field to build up to its maximum

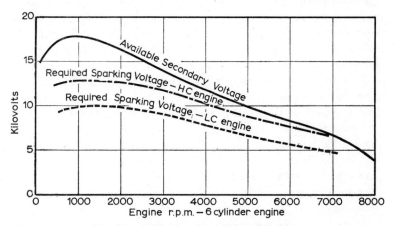

Fig. 16.   Available voltage with coil ignition

value (the saturation time). Despite this, the time available for build-up is still inadequate on the modern high speed engine and the 'available secondary voltage' falls with increase of engine speed. This is illustrated in Fig. 16. The voltage required at the plug gaps to produce the spark increases with cylinder pressure. This required voltage is therefore at its highest value at full throttle and at the engine speed which gives maximum volumetric efficiency. As the volumetric efficiency falls off with increase in speed, so the required sparking voltage also falls. The curve of 'required voltage' (see Fig. 16) will obviously vary from engine to engine, being highest on high compression engines that breathe very well. The broken curve in Fig. 16 shows a typical required voltage curve for a high compression sports car engine. In this case it is seen that the danger of misfiring occurs at speeds above 5000 r.p.m. With this particular engine and ignition system, ignition at high speeds would depend to a very critical degree on the state of tune of the ignition system. It would be advisable in such a case to use a high voltage coil or 'sports coil' to increase the reserve voltage to a satisfactory level.

The phenomena of the spark

The spark that occurs with such apparent ease across the electrodes of a plug is not the simple thing we might believe from the evidence of our eyes. The electrical energy stored in the secondary circuit is discharged in two distinct stages. The first stage, usually called the capacity component, lasts no longer than a hundred-thousandth of a second. The second stage which contains the bulk of the electrical energy, is called the inductive component and lasts for at least one thousandth of a second.

If it were not for the presence of certain electrically charged particles or *ions* in the atmosphere, air would be a perfect insulator and it would be impossible to make a sparking plug work at all. We need not discuss here all the details of *ionisation*, since this is not a scientific treatise. The ionisation of a spark gap can be regarded as a kind of 'breeding' of ions from existing molecules until sufficient ions exist in the gap for the conduction of electricity across the gap. The existing ions collide with neutral molecules. Each disrupted molecule becomes two separate ions with opposite electrical charges, and these, under the action of the magnetic field produced in the spark gap by the voltage across the gap, begin to move in opposite directions, the positive ions moving towards the negative electrode, the negative ions moving towards the positive electrode. These newly created ions produce more and more ions by collision until the gap becomes a conductor of electricity by virtue of the passage of ions between the electrodes. The time interval from the start of ionisation to the instant when the gap becomes fully ionised is extremely small, only the fraction of one thousandth of a second. At this stage the discharge occurs, the capacity component first, followed by the more flaming second component, the inductive part of the discharge which contains the greater part of the energy. The actual *rate* of energy dissipation during the second part of the discharge is not as high as during the passage of the capacity component, since the inductive component is of much greater duration.

With so small a part of the spark energy contained in the initial capacity component one would hardly expect this part of the discharge to contribute much towards ignition of the combustible mixture in the cylinder. Experimental apparatus, in which the inductive component can be reduced in size until it is almost non-existent, has been used to show that satisfactory ignition of a warm engine is given by the capacity component alone. In such conditions with a warm engine and a normal mixture, the inductive component appears to contribute nothing to the process of ignition.

It is on cold mornings, when part of the fuel is still in the form of liquid droplets, that the size of the inductive component is of great importance. The heat input of the capacity component is often insufficient to vaporise enough of the volatile fraction of the petrol and then, having vaporised it, to raise its temperature to the required value for combustion. The heat energy of the inductive component is many times that of the capacity component, and is usually sufficient to ignite a wet cold mixture. The visual effect of the inductive component is much greater than that of the capacity component, the duration of the latter being too short for human perception. We therefore see why the old-time mechanic was never satisfied until a magneto produced a 'fat' spark, since the fatness of the spark is the healthy-looking flame of the inductive component.

There are other times when a fairly large inductive component is of value. With a very weak mixture, combustion is slow and difficult and the flaming heat of the fat spark can sometimes prevent misfiring when weak part-throttle mixtures are used. Again, when a plug is fouled by the products of combustion or oil, the resistance of the plug at its working temperature falls from the normal value of 20-50 megohms to as little as 1 megohm. This low resistance retards the growth of the secondary voltage in the coil. In such circumstances a fat spark can sometimes help to counteract the bad effects of the electrical leakage at the plug.

Sparking plug and contact breaker gaps

We now see how important it is to maintain the gaps between the plug points and the contact breaker points within the prescribed limits. Plug gaps that are too small will not give the fat spark required for cold starting. Plug gaps that are too large, however, raise the general level of the curve of required voltage (Fig. 16) and could lead to misfiring at the top end of the speed range. If the contact breaker points are set to give too wide a gap, too much time is spent in the open position, leaving too little time with the points closed. A drastic reduction in this time does not allow sufficient time for the current build-up in the primary windings and results in a drop in the available secondary voltage curve (Fig. 16).

Dwell angle

The time interval during which the contacts remain closed is usually expressed in terms of degrees of contact breaker cam rotation. When expressed in this form it is called the dwell angle. This angle can be measured by the tach-dwell meter or by the ignition oscilloscope, as described in Chapter Two.

Contact alignment

It is obvious that any reduction in the voltage applied to the primary windings will result in a corresponding percentage drop in the secondary voltage. A voltage drop can be caused by faulty connections, an incorrectly set voltage regulator or, especially when starting, a defective or discharged battery. An excessive voltage drop can also occur at the contact breaker. The points may be dirty or pitted, or they may be in poor alignment, so that the effective contact area is very small. Visual inspection will reveal the more obvious faults, but the use of a low-reading voltmeter to measure the actual voltage drop across the points when closed is the most dependable method of checking the condition of the points. In good condition this voltage drop should not exceed one tenth of a volt.

Condenser

Excessive pitting may be caused by a faulty condenser, by one of inadequate or excessive capacity, or by a condenser which offers too high a resistance. If the 'pit' is on the negative point and the 'mound' on the positive point, the capacity of the condenser could be too low. The reverse is suggested when the mound is on the negative point. A modern instrument for testing condensers was described in Chapter Two.

Ignition timing

Nothing has been said so far about the important business of ignition timing. We have discussed some of the difficulties associated with the production of the spark. It is just as important to produce the spark at the right time as to produce it under all conditions of engine operation. The right time for one engine running condition is not the right time for another. For every combination of engine speed and manifold depression there is one ignition timing, and one only, that gives maximum power. In the early days of motoring the driver was provided with a hand lever which he moved when he wished to advance or retard the ignition timing. He seldom forgot to retard the ignition before trying to start the engine and if he thought the engine appeared to be labouring on a hill he would retard the ignition by what he thought was the appropriate amount. Since driving was such an absorbing occupation forty or fifty years ago there was little danger that the driver would forget to advance the ignition again when the hill was surmounted. Many veteran cars were extremely sensitive to ignition timing and could be 'driven on the ignition lever', rather than by manipulating the throttle.

Since these pioneer times the automobile ignition equipment manufacturers have developed automatic advance mechanisms which are almost perfect in that they reproduce with accuracy the ignition timing advance characteristics specified by the engine manufacturers. All engines require a greater ignition advance at high engine speeds than at low speeds. The time which elapses from the instant of firing the plug to the moment when the flame front has travelled a given distance across the combustion chamber, is influenced by several factors, such as the degree of flame turbulence, the pressure and temperature of the charge and the shape of the combustion chamber. Despite all these different influences there is one factor which overrides all, the speed of the engine itself, since this reduces the available time for combustion. Flame turbulence increases with increase in engine speed, but the increase in flame speed brought about in this way is not sufficient to compensate for the reduction in available time. Thus if combustion is to be completed before the piston has descended too far, the ignition timing (measured in crank-angle degrees before T.D.C.) must be gradually advanced as the engine speed increases. The correct amount of advance (the speed advance curve) is found by dynamometer tests at the maker's works and the ignition component maker designs the weights and springs in the centrifugal advance mechanism to reproduce the required advance curve.

This centrifugal advance mechanism takes no account of throttle opening, being designed simply to reproduce the optimum ignition advance curve for maximum power, i.e. for full throttle running. The centrifugal advance mechanism will give the same advance whether the engine be running at 1500 r.p.m. on the level with a small throttle opening or climbing a hill at 1500 r.p.m. with full throttle. In the former case the compression pressures and temperatures are much lower than in the latter case, the mass flow through the inlet valve is much lower and the overall effect is of a slowing down of the speed of combustion. To compensate for this effect some engines have a second device fitted, the vacuum advance unit, to advance and retard the ignition timing in response to changes in the depression in the induction manifold. The vacuum unit is essentially an enclosed spring-loaded flexible diaphragm which is attached to a shaft which moves the contact breaker baseplate relative to the main body of the distributor. The flexible diaphragm is made to move against the force of the spring by the depression transmitted to the outside face of the diaphragm by means of a length of small-bore tubing connected to a drilling in the carburettor body. This small hole is so placed that it lies on the engine side of the throttle butterfly when the throttle is open. When closed the edge of the throttle covers the hole. Thus when idling, although the maximum induction vacuum is then normally recorded, the vacuum connection is then intentionally covered up and the diaphragm, under the action of the spring, moves the baseplate to the fully retarded position. Idling is a special case, in that the heating effect of exhaust gas dilution becomes so great and the time available for combustion becomes so long, that any ignition advance beyond T.D.C. almost invariably leads to rough running. As the throttle is opened, better exhaust scavenging is achieved, by virtue of the higher cylinder pressures when the exhaust valve opens, less hot exhaust gas is retained to admix with the fresh charge, less time is available for combustion and a greater ignition advance is required. Thus at speeds of 1000 r.p.m. or more, the vacuum drilling in the carburettor body becomes uncovered and the high induction vacuum gives full advance to the distributor.

The two automatic advance mechanisms work entirely independently and their effects are additional in an algebraic sense. Thus it is possible for the driver to increase engine speed, producing

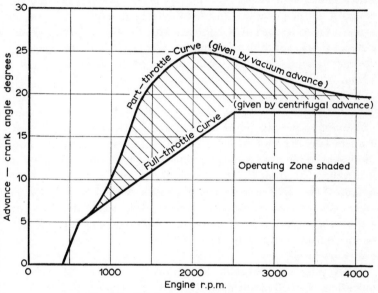

FIG. 17. Typical automatic advance curves

an ignition advance on the centrifugal mechanism, and at the same time, to open the throttle wide enough to cause the induction depression to fall sufficiently for the vacuum advance mechanism to retard the ignition timing by an exactly equal amount. Fig. 17 shows how the ignition timing of the engine is controlled between two limiting curves at all engine speeds, the upper curve, being the part-throttle cruising curve, the lower the full-throttle curve. On a typical town journey the ignition timing will be varying all the time between these limits.

Let us suppose that the vacuum advance diaphragm is leaking, or the centrifugal advance weights are sticking on their pivots. If any defect of this nature is present to interfere with the correct functioning of one or other of these controls, every change in engine speed or throttle-opening will be accompanied by a change to an ignition timing which is incorrect. The efficiency of combustion will be reduced, the fuel consumption will increase, power will be lost and, if excessive advance is sometimes given, detonation or even pre-ignition may occur.

To carry out a complete and accurate check on the behaviour of both ignition controls one must use one of the specialist instruments designed for the purpose. The distributor analyser and the ignition 'scope were described in Chapter Two. Complete failure of the vacuum unit is not difficult to diagnose even without these aids and the methods used to trace such faults are described in Chapter Seven.

Sparking plugs

A sparking plug is simply an insulated spark gap. Nevertheless, it operates under most arduous conditions. It is exposed in rapid succession to high gas pressures, high flame temperatures and then relatively cold draughts of petrol-air mixture. Relatively rich mixtures leave films of carbon on the inside portion of the insulator during the exhaust stroke and general neglect and the dust and grime of modern roads sees a gradual build-up of conductive film on the outside surfaces of the insulator. One wonders how they ever work at all!

Every plug has a working temperature range best suited to its operation. Elevation of the electrode temperature is advantageous since it lowers the sparking potential. It also tends to burn off any carbon deposited on the electrodes, which might otherwise form a conducting track to short-circuit the current. On the other hand the electrode temperature must not be too high, since this might lead to pre-ignition. Since the outer electrode has a good conducting path straight to the plug body and the central electrode is thermally insulated by the electrical insulator, it follows that the central electrode runs at a much higher temperature than the outer. The temperature of this critical central electrode can be controlled largely by its length and diameter. Thus an increase in length makes the plug tip hotter, as does also a decrease in diameter. When operating at the correct temperature the plug tip is in the range 700-750°C (approx. 1300-1400°F).

A plug, as other spark gaps, comprises three elements: two electrodes and the insulator between them. Since they are subjected to very high temperatures and the eroding effects of the spark itself, special materials have been developed to help give long life. For many years plug insulators were made either of mica or porcelain. To-day sintered aluminium oxide is the almost universal choice. It is manufactured under such trade names as Corundite, Sintox, Zircrund and Sinterkorund. The insulator must have adequate dielectric strength to withstand the high voltage between the

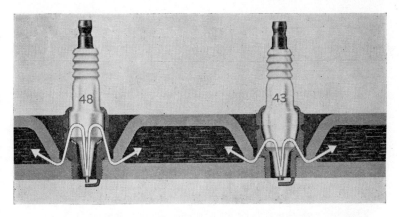

FIG. 18. A hot plug and a cold plug (*Photograph by courtesy AC Spark Plug; The Electronics Division of General Motors*)

electrodes, and must also possess a high figure of insulation resistance. Both of these required properties are affected by temperature, the values decreasing with an increase in temperature. It follows that when insulation tests are made they should be carried out at the working temperature of the plug. The value of the insulation resistance is also a function of the test voltage, falling as the voltage rises. Plug resistance tests should therefore be carried out at the highest working voltage.

Fig. 18 shows two A.C. plugs of different heat value. Heat picked up by the insulator tip by convection and radiation from the burning gases can only be conducted away vertically along the

SCE E

body of the insulator to the point of maximum cross-section, from which point the heat can flow outwards into the steel body and thence to the surrounding air and to the water-cooled metal of the cylinder head. It will be appreciated that the temperature of the insulator, and consequently of the central electrode, depends upon the distance between the insulator nose and the seat in the body. Cold plugs have a relatively low seat, hot plugs a high one. The former plugs will be used in a hot-running engine, the latter type in a cold-running one. The temperature range specified earlier is determined by two things, carbon-formation and pre-ignition. Below 550°C (1000°F) carbon deposits do not readily burn off the plug tip. Above 850°C (1550°F) the plug tip becomes so hot that it can start pre-ignition.

The electrode material must naturally be such that it resists erosion from the spark and corrosion from the products of combustion. For most automotive sparking plugs the electrodes are made of a nickel alloy, usually with a small percentage of manganese. For central electrodes, silicon-manganese-nickel alloys are used, since these are resistant to the high temperature attack from sulphur and T.E.L. in the fuel. Platinum alloys are far more resistant to attack and are also less prone to induce pre-ignition. Despite the higher cost they are popular for high-performance sports cars on this account.

MAGNETOS

The inherent advantage of magneto ignition over coil ignition as a good spark producer at high engine speeds has been slowly whittled away by slow steady improvements in the design of coils. Consequently the advantages of simplicity and cheapness have made coil ignition the almost universal system. On racing cars, where a battery is not required for starting, lighting and many other auxiliary services, the weight penalty of the coil system with its attendant battery becomes a serious factor and magneto ignition has remained favourite in the field.

The principle of the magneto

All magnetos, whatever the configuration, have the following essential components:

(*a*) magnet(s)

(*b*) primary and secondary windings on an iron core

(*c*) a contact breaker

(*d*) a high tension distributor.

From this it would appear that a magneto is simply a coil ignition system with the addition of a rotating magnetic field to supply the

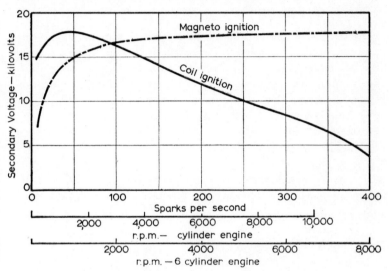

Fig. 19. Comparison of coil and magneto ignition

electromotive force. This may be an oversimplification, but it does help to show the relationship between the two systems.

Classes of magneto

(*a*) *Rotating armature:* This is one of the earliest types. An H-section armature, carrying primary and secondary windings, rotates between the pole pieces of a magnet.

(*b*) *Rotating magnet:* A disadvantage of the above type is that the windings are subjected to very high centrifugal loadings, especially on very high speed engines. By rotating the magnet and fixing the armature a more robust structure is possible. This is the principle of the Scintilla Vertex magneto shown in Fig. 20.

(c) *Sleeve inductor:* In this type the H-section armature is stationary, the magnet is stationary and a sleeve inductor rotates in
the magnetic field between the armature and the magnet
pole pieces. Gaps in the sleeve interrupt the magnetic field
to produce the required changes in flux density.

(d) *Polar inductor:* This design has a rotating inductor with an
even number of poles. These interrupt the magnetic field of
the fixed magnet.

All the above configurations have a common aim, i.e. the production of a cyclical variation in the magnetic flux that passes through
the laminated iron core on which the primary and secondary turns
are wound. All magnetos differ in one respect from coil ignition.
In coil ignition, when the contacts open the flux in the core of the
coil drops to zero. In a magneto, not only does it drop to zero at a
greater rate, but it drops past zero until the same flux density is
reached in the opposite direction.

It is so rare to-day for magneto ignition to be used on a sports
car that the maintenance of magnetos is now largely left in the
hands of specialists. Moreover, it is a breed that is rapidly dying
out.

Care and maintenance

Internal timing. The internal timing of a magneto only concerns
us when it has been fully dismantled for overhaul. It concerns the
relative positions of the main rotating member, the contacts and
the distributor brush or spark-gap. Since magnetos differ so much
from design to design no general instruction can be given here. In
any case it would be inadvisable to strip a magneto without the
assistance of the maker's service manual.

External timing. This operation is essentially the same as the
action of timing a distributor in a coil ignition system. A vernier
coupling or other timing device is usually placed in the drive to the
magneto to permit small changes in timing to be made until correct
to specification.

Periodic maintenance

Lubrication. One should avoid over-lubrication. A few drops of
oil supplied to rotor or gear-wheel bearings every 5000 miles is

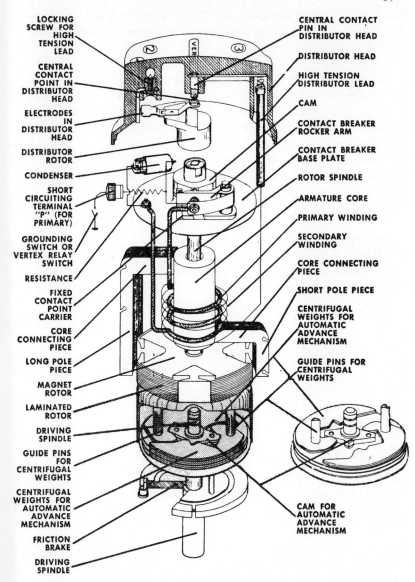

LOCKING SCREW FOR HIGH TENSION LEAD

CENTRAL CONTACT POINT IN DISTRIBUTOR HEAD

ELECTRODES IN DISTRIBUTOR HEAD

DISTRIBUTOR ROTOR

CONDENSER

SHORT CIRCUITING TERMINAL "P" (FOR PRIMARY)

GROUNDING SWITCH OR VERTEX RELAY SWITCH

RESISTANCE

FIXED CONTACT POINT CARRIER

CORE CONNECTING PIECE

LONG POLE PIECE

MAGNET ROTOR

LAMINATED ROTOR

DRIVING SPINDLE

GUIDE PINS FOR CENTRIFUGAL WEIGHTS

CENTRIFUGAL WEIGHTS FOR AUTOMATIC ADVANCE MECHANISM

FRICTION BRAKE

DRIVING SPINDLE

CENTRAL CONTACT PIN IN DISTRIBUTOR HEAD

DISTRIBUTOR HEAD

HIGH TENSION DISTRIBUTOR LEAD

CAM

CONTACT BREAKER ROCKER ARM

CONTACT BREAKER BASE PLATE

ROTOR SPINDLE

ARMATURE CORE

PRIMARY WINDING

SECONDARY WINDING

CORE CONNECTING PIECE

SHORT POLE PIECE

CENTRIFUGAL WEIGHTS FOR AUTOMATIC ADVANCE MECHANISM

GUIDE PINS FOR CENTRIFUGAL WEIGHTS

CAM FOR AUTOMATIC ADVANCE MECHANISM

Fig. 20. Exploded view of Scintilla Vertex magneto which operates on the rotating magneto-stationary coil principle. The advance mechanism is centrifugally controlled with weights against the resistance of the breaker cam. By using different weights the advance curve can be adapted to any type engine.

usually all that is necessary. At the same time contacts should be re-faced and re-set to the recommended clearance and the distributor housing wiped clean.

The Scintilla Vertex magneto

This magneto has achieved such world-wide fame for its reliability and excellent performance that a brief description is warranted. It has been a popular tuning fitment for many years on small European sports cars. Fig. 20 is an exploded view with all essential components labelled. An automatic centrifugal advance mechanism is housed in the base. A wide range of balance weights is available from the factory to give any desired advance curve between 0° and 70°.

New developments

Recent innovations in the field of automotive ignition, such as transistorised systems, electronic systems and the spark-pump are described in Chapter Eleven.

CHAPTER FOUR

Mixture Formation, Distribution and Carburation

The mixture

Commercially available petrols are a complex mixture of hydrocarbons. The chemical differences between the separate hydrocarbons influence the combustion behaviour. Even when we consider hydrocarbons with the same number of carbon and hydrogen atoms in their molecules we can still find that their combustion behaviour is different, for the way in which the carbon and hydrogen atoms are linked together within the structure of the molecule has a profound influence on the manner in which it will burn. For example, *normal*-octane and *iso*-octane have the same chemical formula, C_8H_{18}. *Normal*-octane has the atoms arranged

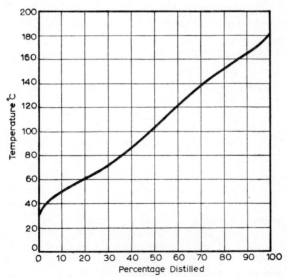

FIG. 21.   A.S.T.M. distillation curve

59

in a pattern usually called 'straight-chains' and the resistance to knocking combustion is very low. *Iso*-octane has the atoms arranged in a more compact form, called 'branched-chains', and the resistance to knocking combustion is very high.

The boiling point range

An important physical property of a petrol is the boiling point range. A fuel made up entirely of the same hydrocarbon, such as *iso*-octane or benzene, will all boil off at the same temperature in the manner of water. In the case of *iso*-octane the boiling point is 99°C, almost the same as that of water. A well-balanced motor fuel will have a carefully blended range of boiling points, starting with a small percentage of butane, which boils at 40°C and finishing with a small amount of low volatility fuel with a boiling point approaching 200°C. A typical distillation curve is given in Fig. 21.

Hydrocarbons are usually divided into four main groups:

1. Paraffins, with the general formula C_nH_{2n+2}
2. Naphthenes, with the general formula C_nH_{2n}
3. Olefins, with the general formula C_nH_{2n}
4. Aromatics, with the general formula C_nH_{2n-6}

When a hydrocarbon burns in air the carbon combines with oxygen to form carbon dioxide (CO_2) and when insufficient oxygen is available or combustion is incomplete for other reasons, a quantity of carbon monoxide (CO). The hydrogen combines with oxygen to form water vapour (H_2O). From the chemical formula of a particular hydrocarbon we can calculate how much air is required

TABLE 2

Compound	Formula	Group	B.P. °C	Chemically correct air/fuel ratio, by weight
Hexane	C_6H_{14}	Paraffin		15·2
Octane	C_8H_{18}	Paraffin	99	15·0
Di-isobutylene	C_8H_{16}	Olefin	104	14·9
Toluene	C_7H_8	Aromatic	110	13·6
Benzene	C_6H_6	Aromatic	80	13·4
Ethyl alcohol	C_2H_6O	Alcohol	78	8·9
Methyl alcohol	CH_4O	Alcohol	65	6·5

to give complete combustion of all the carbon and all the hydrogen. This can be expressed as an air/fuel ratio, so many pounds of air to so many pounds of fuel. This is called the chemically correct or the stoichiometric mixture. Since the proportions of carbon to hydrogen differ in different hydrocarbons the chemically correct air/fuel ratio also differs. Several examples are given in Table 2. Methyl and ethyl alcohol are included for comparison.

Despite the wide variation shown above in the chemically correct air/fuel ratios of many common fuel components the commercial motor fuels sold to-day are all blended to conform to a very narrow specification. Even though they differ appreciably in aromatic content, even in knock rating, the chemically correct air/fuel ratio falls into the bracket 14·5-14·9 to 1. With very little chance of error one can take a mean value of 14·7 to 1 as the chemically correct air/fuel ratio for all petrols sold at the pump to-day. Special blends of racing fuel containing substantial quantities of alcohol will obviously have a much lower correct air/fuel ratio.

Mixture and power

The chemically correct mixture strength does not give maximum power; neither does it, in general, give maximum economy. Fig. 22 shows what is called a mixture loop or 'fish-hook'. By running an engine at full-throttle and maintaining a constant speed on the dynamometer the power developed at different mixture strengths can be measured. A special adjustable carburettor is required to permit the wide variation in mixtures. Fig. 22 is a plot of specific fuel consumption, in pounds of fuel per brake horse power per hour, against brake mean effective pressure. This latter quantity, usually abbreviated to b.m.e.p., is the mean pressure exerted on the piston during the working stroke in terms of measured torque at the flywheel. Indicated mean effective pressure is measured directly at the cylinder and is greater than the b.m.e.p. by the amount of effective pressure lost in internal friction in the engine. It is of value to the practical engine tuner to regard an increase in b.m.e.p. as equivalent to an increase in torque. The torque of the engine is in fact proportional to the b.m.e.p. times the swept volume of the engine. From Fig. 22 we see that for a typical engine maximum power is given at an air/fuel ratio of about

12·5 to 1. Maximum economy is given at about 17 to 1. At weaker ratios than 17 to 1 it is usually difficult to get the engine to run at a steady speed. The engine will 'hunt' in speed, i.e. rise and fall. It will be readily seen from this that some power or economy

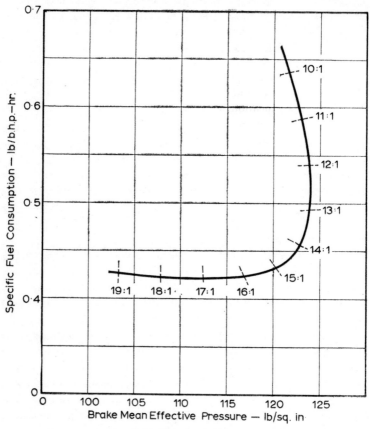

FIG. 22.   A typical mixture loop

(whichever happens to be our target) can be lost if the mixture distribution between cylinders is poor. For example, a four cylinder engine with two cylinders receiving a mixture of 11 to 1 and the other two receiving a mixture of 14 to 1 will have an average mixture strength of 12·5 to 1 but the power developed will be 3 or 4 per cent down on an engine with perfect distribution.

Distribution

A study of Fig. 22 shows us that mixture distribution between cylinders need not be held to extremely narrow tolerances. Only very bad discrepancies result in much loss of power. Were it imperative to maintain mixture strengths between cylinders to a tolerance of plus or minus half an air/fuel ratio there would have been a much greater incentive in favour of fuel injection direct into the cylinder. Having given the impression that 'anything goes' so long as we get some of the fuel into some of the cylinders, let us correct this with a practical example. Some V-8 engines made in recent years have such bad distribution when running light, especially after a cold start, that the mixture cannot be adjusted to give a combustible mixture to all cylinders. This means that the range of mixtures between cylinders spans about 6 air/fuel ratios. There is no cause for complacency there.

Formation of the mixture

The detail design of modern carburettors and the methods used by carburettor designers to maintain the desired air/fuel ratios over the whole range of engine operations are described later in this chapter and in the two succeeding chapters. Let us assume then that the problems of metering the fuel into the air stream will be solved later. Our concern at the moment is the behaviour of this spray of liquid fuel after it leaves the venturi.

At all but the slowest speeds the liquid fuel spray that enters the air stream at the carburettor venturi is in a very finely divided form. Very little of it is in vapour form at this stage. As they pass the edge of the throttle-plate, however, the fuel droplets enter a zone of changed ambient conditions, and depending upon the prevailing manifold vacuum, a process of fairly rapid evaporation begins. On a typical sports car engine at a road-load of 50 m.p.h. the induction vacuum will be about 15″ Hg or half an atmosphere. Thus the droplets of petrol, as they pass the edge of the throttle-plate, will pass from a pressure zone of 30″ Hg to one of 15″ Hg. The more volatile fractions of the mixture will flash off into vapour immediately, the middle fraction will evaporate more slowly as heat is picked up from the manifold and the 'heavy end' will still be in liquid form as the mixture enters the cylinder. A stabilising influence on the amount of evaporation occurring after the throttle-plate is the limit set by the rate at which heat can flow into the manifold, whether it be from a hot spot, a water jacket, or from the unheated outer surface of the manifold. The latent heat of a typical petrol is 75 C.H.U. per lb. (135 B.T.U. per lb.) A refrigeration effect is inevitable if the manifold is unheated. This causes the undesirable effects of snow and ice formation on the unheated throttle-plate that we read so much about in the petrol company advertisements. It also acts as a deterrent to the evaporation of mixture. On a cold engine it is almost impossible to get enough heat into the manifold. When the engine is warm, the amount of heat given to the manifold must be carefully controlled by good design or power will be wasted by decreasing the density of the charge.

Except for the case of a cold engine, vaporisation is usually complete before the inlet valve closes. The piston crown, cylinder walls and especially the exhaust valve head present hot surfaces to the turbulent mixture as it swirls into the cylinder. Moreover the hot residual gases from the previous cycle are mixed thoroughly with the new charge during the induction process.

Manifold design

The design of an induction manifold would be relatively simple if all the mixture were vaporised at the throttle-plate. At light loads about one third of the mixture handled by the manifold is in a liquid state and at low induction vacuums—6" Hg or less, about half of the mixture passing around the bends in the manifold is in a wet state. The more bends to negotiate, the more difficult the problem. With a racing engine having one carburettor per cylinder and a nearly-straight passage all the way from venturi to valve wet mixtures are no problem at all. This despite the poor atomisation given by the use of such large venturis. The walls are wet with fuel most of the time, but this in itself will not upset distribution. When several cylinders share the mixture from one carburettor the behaviour of the mixture in the manifold becomes of primary importance. The reader will be familiar with the customary right-angled bends used on the modern manifold. Such a design is a compromise between the conflicting requirements of air flow and fuel flow. If it were not for the presence of liquid fuel more streamlined passages would be used. Fig. 23 shows diagrammatically

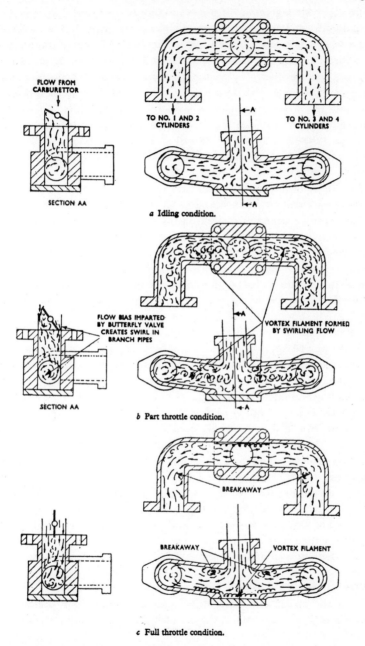

FLOW FROM CARBURETTOR

TO NO. 1 AND 2 CYLINDERS

TO NO. 3 AND 4 CYLINDERS

SECTION AA

a Idling condition.

FLOW BIAS IMPARTED BY BUTTERFLY VALVE CREATES SWIRL IN BRANCH PIPES

VORTEX FILAMENT FORMED BY SWIRLING FLOW

SECTION AA

b Part throttle condition.

BREAKAWAY

BREAKAWAY

VORTEX FILAMENT

c Full throttle condition.

FIG. 23. Flow patterns in 4-cylinder manifold (*From Paper read to Institution of Mechanical Engineers in Feb. 1961, by J. S. Clarke, C.B.E., B.Sc., Ph.D.*)

the observed behaviour of a wet mixture in a typical four-cylinder manifold. A Perspex (transparent plastic) manifold was used for the tests which were carried out in the Joseph Lucas Laboratories in Birmingham. As the air flow separates into two channels the larger fuel droplets, by virtue of their inertia, fail to turn the corner. Where they strike the walls is the usual place to incorporate a well and a hot-spot to help evaporate this wet fuel. Once the manifold walls become wet with fuel it becomes difficult to re-entrain this fuel into the gas stream. With smooth walls and sweeping bends this film would often persist all the way to the valve seat. The provision of a sharp edge to the inner part of a bend helps to shoot much of this fuel back into the gas stream. This indi-cated by 'breakaway' in Fig. 23. George Mangoletsi of the G.M. Carburettor Company carried out experiments before the last war that convinced him that a right-angled bend is only partially successful as a re-entrainment device. Experiments with knife-edged annular projections fitted between the carburettor flange and the manifold led to the development of the famous G.M. Modifier and the finely engineered range of modified induction manifolds supplied by this company.

Experience has shown that the amount of wet fuel on the manifold walls must be kept to a minimum if good snap accelerations are to be achieved without 'fluffing'. When the engine is at operating temperature the amount of fuel retained on the manifold walls varies with manifold depression. With a high depression the amount is relatively low. Thus after a period of high vacuum running, such as a spell of 30 m.p.h. traffic crawling, the walls have dried out. If then the throttle is suddenly snapped open the fuel leaving the venturi enters a manifold in which there is too low a vacuum to produce much evaporation. The dry manifold walls thus act like a sponge during this initial acceleration period and, until the walls reach their equilibrium wetted condition the fuel/air stream is robbed of fuel. The provision of additional fuel from an accelerator pump is the popular solution to this difficulty, but the reduction of the manifold wall area can make this less of a problem.

Pressure losses by distribution

To the sports car tuner the induction manifold represents a source of power loss. Thus the use of sharp bends conflicts with our primary object of inducing the maximum possible charge into the cylinders. The problem of mixture distribution between cylinders becomes less of a problem as more carburettors are used and there is a tendency to-day to streamline the bends more on twin-car-

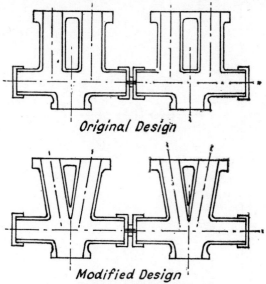

Original Design

Modified Design

FIG. 24.   Induction systems on FWA Type
Coventry-Climax engine

burettor 4s, using Y-junctions instead of Ts. When such a change was made to the 1100 c.c. single o.h.c. Coventry Climax engine (see Fig. 24) the maximum power increased by 3 b.h.p.

One of the sources of power loss in a four-cylinder manifold is the reversal of flow that occurs twice per cycle on a single car-burettor manifold, such as used on many of the B.M.C. engines. These engines have siamesed ports to Nos. 1 and 2 and to Nos. 3 and 4 inlet valves. The writer designed his 'Uniflow' manifold for the A Series engine to overcome this problem. It produced 5 extra b.h.p. from the standard A40 engine and lost nothing in m.p.g. When a similar design was fitted to a Ford Prefect engine a substantial improvement in m.p.g. and power was given. It will be seen from Fig. 25 that the flow in the doughnut-shaped manifold is uni-directional, thus keeping the gas moving towards the valves at all times. On the standard manifold the column of gas has to change directions between the induction periods of Nos. 1 and 3 and between Nos. 4 and 2. The energy to achieve these direction changes can only come from the total energy of the gas stream.

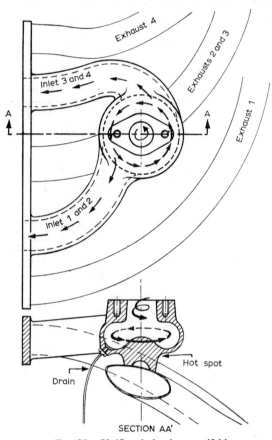

FIG. 25.   Uniflow induction manifold

Thus the absolute pressure of the charge entering the cylinder is unnecessarily reduced.

THE MIXTURE REQUIREMENTS OF THE ENGINE

The engine is the customer—and the customer is always right. It is a common misconception that the perfect carburettor would feed the same mixture strength to the engine at all times—at all engine speeds and at all throttle openings. Of course, every motorist knows that a richer mixture is required to start a cold engine, but in general it would seen to the layman that this is a special case and at all times when the choke is not in use the mixture provided by the carburettor should be constant. This is not so and the mixtures demanded by our engine for regular running can range from 12 to 1 (twelve pounds of air with every pound of fuel) in some cases to as weak as 16 to 1 at other times. For maximum economy it is necessary to design our carburettor to take full advantage of all occasions when the engine will burn weak mixtures (and when such mixtures may be used without harm). It is customary to consider the mixture requirements of an engine under five general headings. These are: starting (and cold-running); idling; part-throttle running; full-throttle running; and the special (and frequent) case of accelerating.

Starting and cold-running

To start an engine from cold it is necessary to provide a very rich mixture at the carburettor. This is necessary to ensure that a mixture of ignitable proportions reaches the plug points. Only the lighter fractions of the petrol blend can be evaporated when the induction tract, cylinder head and cylinder walls are near freezing point. In England winter starting calls for a mixture strength of about one pound of fuel to four pounds of air. In colder climates a mixture as rich as 1 to 1 is sometimes necessary to obtain a start. Until the engine reaches working temperature a richer-than-normal mixture must be provided. The degree of enrichment may be controlled by some automatic device such as the auxiliary enrichment carburettor on the Jaguar or the automatic strangler fitted to some Stromberg carburettors. In the majority of cases on British cars the mixture control is manually operated, and oddly enough, this seems to lead to greater economy.

Idling

When an engine is idling, several factors combine to hinder good mixture distribution and combustion. Only one factor—the high depression in the induction system—assists fuel vaporisation. The low mass-velocities in the induction tract and the poor turbulence

SCE F

at the valve throats tend to give poor vaporisation and poor mixture distribution. The slow piston speed reduces turbulence and retards the speed of combustion. On the modern engine valve timings are chosen to give good breathing at high engine speeds. The customary valve overlap of 20 to 30 degrees at top dead centre can lead to a back-flow of exhaust gas into the induction tract when the inlet valve opens. On engines with sports/racing camshafts, giving overlaps of 60 to 80 degrees, good idling is sometimes an impossibility below a speed of 1000 r.p.m. The shape of the combustion chamber and the degree of swirl turbulence given by the inlet valve both have a profound influence on satisfactory slow-speed combustion. In general, however, it is necessary to provide a rich mixture at the carburettor to compensate for the poor distribution and vaporisation. In this way the mixture given to the cylinder receiving the weakest charge is still ignitable. The mixture received by the cylinder that is favoured by mixture-bias must not be too rich for combustion. Some engines require a mixture as rich as 12 to 1 to give smooth combustion, but the average will be nearer 13 or 13·5 to 1.

Part-throttle operation

As the throttle is opened and the engine speed increases above 1000 r.p.m., satisfactory combustion can be achieved over a wide range of mixture strengths. As we reach a throttle opening giving approximately 40 per cent of full power at the particular engine speed, it is possible to burn mixtures ranging from 10 to 1 down to 16 to 1. The limits can be extended by careful attention to cylinder head design and swirl-turbulence induced at the inlet ports.

Maximum power *at a given throttle setting* requires a mixture strength of between 12 and 13 to 1. If we maintain this throttle setting and weaken the mixture to 16 to 1, the power will drop by about 10 per cent. This is obviously more economical since 90 per cent of the maximum power given at this setting is given by only 70 per cent of the maximum-power fuel consumption. If, then, the throttle is opened slightly to recover the original power petrol is saved, since 90 per cent of the maximum power setting can be achieved. This, in brief, is the method used to achieve economy on the modern carburettor. It can be achieved in this way, without the driver being at all conscious of the fact.

Full throttle

The danger of valve-burning under full-throttle temperatures is very much increased by the use of part-throttle economy mixtures. In general it is inadvisable to use mixtures weaker than 13·5 to 1 at full throttle. In theory, of course, serious oxidation should not occur until the mixture is weaker than the chemically correct mixture ratio of approximately 14·5 to 1. (This value varies slightly depending upon the composition of the petrol.) In fact, however, the combination of full-throttle valve head temperatures of about 700°C (about 1290°F) and mixture strengths weaker than 14 to 1 can prove too severe a test for the normal exhaust valve steel and burning of the valves' mitred seat can occur.

Fig. 26 has been constructed to illustrate the mixture requirements of a typical modern engine. Separate curves are drawn for three engine speeds. These illustrate how the mixture requirements

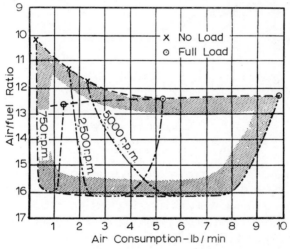

FIG. 26. The mixture 'envelope'

change as the air consumption increases from no-load up to full-load conditions. Even at high r.p.m. a rich mixture is required to promote regular combustion under no-load (overrun) conditions. As the throttle is opened, weaker mixtures can be burned and over most of the part-throttle range a mixture strength of 16 to 1 can be used. As we approach full-throttle, from about 80 per cent of full-

throttle air consumption, it is necessary to provide a gradual enrichment up to the full-throttle mixture of 12-13 to 1. This figure
thus gives us the limiting 'envelope' of mixtures, the shaded area
enclosing the limiting range of mixture strengths. Thus to take a
particular case, at 5000 r.p.m. and an air consumption of 6 lb. per
minute the particular engine will run satisfactorily and efficiently
on a mixture of 16 to 1. It will also run satisfactorily on richer
mixtures up to as rich a mixture as 12·5 to 1. but this would
patently be very wasteful. With the weaker mixture the throttle
opening would be slightly greater, of course, but this in itself is

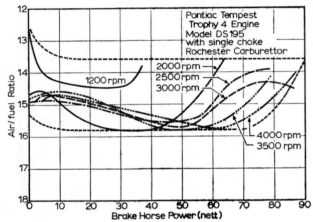

FIG. 27.   Curves of air/fuel ratio against horsepower.
Pontiac Tempest 'Trophy 4' engine

not detrimental. If therefore we are to achieve the utmost economy, the part-throttle economy device must be specially tuned to
match the extreme weak mixture limits of the engine. Curves of
air/fuel ratio against horse power, measured by the writer on a
Pontiac Tempest 'Trophy-4' engine, are given in Fig. 27.

Acceleration

One special condition of engine operation remains to be considered, the case of acceleration. When the throttle is opened
suddenly, the depression in the induction tract drops to a very low
value only recovering and finding a new equilibrium value after
the engine speed is steady again. If no special provision is made in
the design of the carburettor, the start of this period of acceleration

shows a marked tendency for the engine to hesitate or misfire. It is customary in textbooks on the subject to blame this entirely on the inertia of the fuel metering system which makes the fuel fed to the jets lag behind the increase in air rate. Another factor, which must of course vary with manifold design, is the tendency for fuel to accumulate on the walls of the manifold. Either of these effects would result in the mixture reaching the cylinders being weaker than the mixture supplied by the carburettor. Thus to ensure a normal mixture at the plug points we would supply a rich mixture at the carburettor. From measurements made with the aid of an exhaust gas analyser the writer has observed that the maintenance of a normal mixture in the cylinder is not, in itself, sufficient to prevent the hiatus. The analyser shows that a *richer-than-normal* mixture is required to give smooth acceleration.

Poor fuel atomisation and incomplete vaporisation of all the droplets before combustion is probably responsible for this inability to burn a normal mixture. The combination of a wide-open throttle and a low engine speed gives poor fuel atomisation and low inlet turbulence hinders vaporisation of the relatively large fuel droplets. Whatever the cause, there is no doubt of the effect and it is customary to make some provision, especially in carburettors for high performance engines, for an automatic enrichment to occur during acceleration.

CHAPTER FIVE

Variable Choke Carburettors

THE S.U. CARBURETTOR

THE working principle of the latest design of S.U. carburettor is identical to that of the carburettors used on the Bentleys at Le Mans more than thirty years ago. This working principle must surely be known to the majority of motoring enthusiasts by now and a brief description of the operation of the S.U. carburettor is all that is called for here.

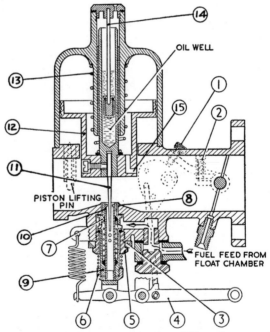

FIG. 28. Section through a modern S.U. carburettor. In these instruments the weight of the piston is, as shown, augmented by a light spring.

At idling engine speeds the base of the piston (part 12 in Fig. 28) rests on the bridge (part 8), two small protuberances on the base of the piston preventing complete stoppage of the air supply to the engine in this position. The depression created downstream of the piston is in communication at all times with the upper part of the piston, the suction disc, by means of the small passage indicated at 15 in Fig. 28. As the throttle is opened the air flow through the gap below the piston increases and the depression downstream of the piston (and above the suction disc) increases until the upward force exerted is exactly equal to the weight of the piston. Any increase in throttle opening from this point causes the piston to lift until an equilibrium position is reached. This equilibrium occurs when the increase in choke area (the area at the bridge) has reduced the depression above the vacuum disc to the point where it just balances the weight of the piston assembly. On the modern design the weight of the piston is augmented slightly by a light spring (13 in Fig. 28). As the piston rises and falls under the influence of the varying demands of the engine so the position of the tapered needle (11) varies in the petrol jet (10), the annular space between the needle and the jet being smaller for low air rates than for high. By choosing the profile of the needle carefully almost any range of mixture strengths can be metered. The shape of the choke, in effect the passage between the needle and the jet, being smaller for low air rates constitutes a rather one-sided venturi and a depression is created at the narrowest part to induce a flow of petrol at the jet. This depression is maintained at a fairly constant level by the movement of the piston. The usual designed value is eight inches head of water (about eleven in. head of petrol). The strength of the piston spring modifies this value slightly to suit individual cases.

In Chapter Four we considered the mixture requirements of the engine under five headings: starting (and cold-running), idling; cruising, full-throttle operation and accelerating. Let us examine the S.U. carburettor and see what special provisions have been made to satisfy these requirements.

Starting

The jet (10), can be moved vertically between an upper jet bearing (7) and a lower jet bearing (9). The enlarged head of the jet is held up to the jet adjusting nut (6) under the action of the spring at the end of the jet lever (4). To provide a rich mixture for starting or cold-running the jet lever (4), which pivots about the centre clevis pin, is pulled upwards by the mixture control cable, thus lowering the position of the jet relative to the needle. In its normal idling position the clearance between the needle and the jet is extremely small, usually about 0·001 in. By lowering the jet $\frac{1}{16}$ in. the clearance will be increased to 0·0025 in. on a typical needle, thus more than doubling the annular metering area and increasing the mixture strength from about 12 to 1 to approximately 5 to 1.

On a few of the larger engines, such as the Jaguar, an electrically-operated auxiliary carburettor is used for starting. This starter carburettor is brought into action when a solenoid is energised by the current from a thermo-switch mounted on the water-jacketed induction manifold. At manifold temperatures above 35°C the thermo-switch opens and the starter carburettor goes out of action. A useful modification for the XK 120 and XK 140 Jaguars is an overriding illuminated switch on the instrument panel. By switching off the starter carburettor manually a considerable saving in fuel can be achieved by a canny driver. If the driver forgets to switch off, the thermo-switch will eventually do it for him. This manual switch is now standard on the later models of Jaguar.

Idling

With a fixed-choke carburettor, such as the Solex or Zenith, the main metering jets admit fuel to the air stream at the choke, as in the S.U. carburettor. At idling air speeds, however, the depression created inside the choke of a fixed-choke carburettor is insufficient to atomise the petrol and a separate jet system is arranged to meter petrol to an orifice downstream of the throttle-edge, where the depression is high. This complication is unnecessary in the case of the S.U. carburettor, since the reduction in choke area at low air-rates maintains a fairly high depression and good atomisation is assured at all times.

Full throttle and cruising

Although a part-throttle economy device has been developed for the S.U. carburettor, this fitting is seldom found to be necessary.

In general part-throttle economy comes from a natural phenomenon exhibited by the engine itself. It is well known that the intermittent nature of the induction process results in a wave formation in the induction tract. At full throttle the petrol jet is subjected to the full effect of these pulsations. A half-closed throttle plate acts as a partial screen against these pulsations and a more even air flow occurs across the jet under part-throttle conditions. By a happy chance a pulsating air flow with a mean air rate of, say, x lb. per minute induces a greater fuel flow from a jet than a steady air rate of x lb. per minute. Thus a full-throttle consumption of x lb. per minute at 3000 r.p.m. will result in a richer mixture than a part-throttle consumption of x lb. per minute at 5000 r.p.m. The pulsations are usually of greater magnitude on a four-cylinder engine than a six. Expressed in the jargon of the carburettor technician the four-cylinder engine gives a greater *mixture ratio spread* than the six. For such engines as the three-litre Rover, where the

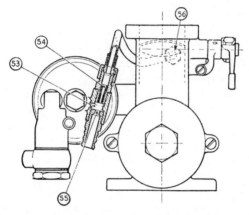

FIG. 29. Showing the device fitted to some current carburettors where the mixture ratio spread is inadequate for the particular engine.

spread is inadequate, the device shown in Fig. 29 is fitted to the float chamber cap nut. The interior of the fitting (53 in Fig. 22) is in communication by means of drillings with the air space above the float, this being the only vent to the float chamber from atmosphere. One side of the fitting is connected to a drilling at the throttle edge (56) the other side being open to atmosphere through a jet (55). With the engine idling, the drilling is on the carburettor side of the throttle plate and the depression transmitted to the device is only the choke depression of about 8 in. water-head. This has a negligible depressive effect on the pressure above the petrol in the float chamber. At full-throttle too, the depression is low and the effect is negligible. With part-throttle conditions, and the throttle plate opened sufficiently for the drilling to be downstream of the throttle edge, a high air rate is induced through the device and the pressure in (53) and in the interior of the float chamber falls appreciably below atmospheric. This reduces the difference in pressure across the jet, and the mixture strength is correspondingly reduced. To limit the extent of the depression applied to the float chamber and to prevent over-weakening of the mixture a venturi restrictor is placed at the pipe inlet (54). The use of the device on carburettors not originally provided with it is not, however, recommended by the makers.

Acceleration

The original undamped S.U. carburettor was regarded generally as a 'power producer', but not as an economical device. The provision of a damper to the piston has removed this disadvantage. A certain amount of enrichment during acceleration can be given on any carburettor by placing the float chamber in front of the jets, thus augmenting the gravitational head by an additional thrust from the inertia of the fuel in the bowl. This additional head unfortunately is insufficient for the purpose and, on the undamped S.U. carburettor, the general mixture level had to be enriched for all conditions to prevent a flat-spot during snap acceleration. The damper fitted to the modern S.U. carburettor (at the base of the spindle, part 14 in Fig. 28) is in effect a floating sleeve working inside the hollow guide of the main carburettor piston, part 12. This sleeve is held rather loosely between a neoprene-seated disc valve above it, and a circlip (or press-stud) below it; these parts are attached to the long spindle, part 14, screwed into the damper filler cap. Without the damper, snap-opening of the throttle can send the piston rocketting to the top of the suction chamber, momentarily dropping the depression at the jet and over-weakening the mixture. With the oil-well charged with the light oil (S.A.E. 20) the lifting of the piston forces oil past the sleeve and the oil drag presses the sleeve up against the disc valve. From this stage the only passage open to the flow of oil is the small clearance between the sleeve and the walls of the piston guide. This reduces the rate of rise of the piston, and by enhancing the depression at the jet, provides a momentary enrichment during the period of acceleration. The resistance to the flow of oil in the opposite direction is negligible, since the circlip (or press-stud) provides free drainage.

The HD type of diaphragm-jet-type carburettor

On the older H type carburettor a potential source of petrol leakage has always existed where the jet tube slides inside the upper and lower bearings. The fitting of new seals effects a cure,

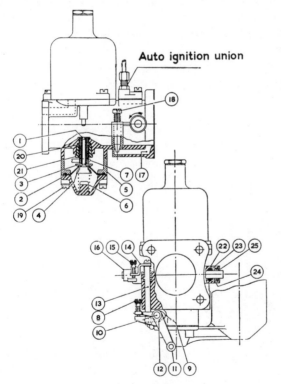

Fig. 30.   The HD-type S.U.
carburettor

but the S.U. Carburettor Co. Ltd. have been well aware of this defect and the HD type was a complete re-design of the jet petrol-feed arrangements to remove this irritating drip.

On the H type petrol enters the jet tube through holes in the side; on the HD type fuel is admitted from below. Vertical movement of the jet (Part 1 in Fig. 30) can take place inside a tubular full-length bearing (20). The movement of the jet is controlled by a return spring (6) and the actuating lever (7). There is no need for jet glands in this design and the only sources of leakage from the jet feed system are by perforation of the synthetic rubber diaphragm (2) or an ineffective seal on the outer flanged joint of the diaphragm cover. A second novel feature on this carburettor is the provision of a passage to by-pass the throttle-plate. The idling mixture is still metered by the single main jet but the throttle is completely closed and the quantity of mixture by-passing the throttle by means of passage 17 is metered by the 'slow-run valve' 18. A much more sensitive control of idling speed is given in this way.

Starting enrichment is given by movement of the jet control lever 11, which is attached through spindle 10 to the jet actuating lever 7. Fast idling during warm-up is given by an interconnection of mixture control and throttle-stop. Movement of lever 11 causes cam 9 to push down the cam-shoe 12, which is attached to the push-rod 13. This opens the throttle by pulling down the throttle-stop 14. Adjustment of the two screws, 8 and 15, will provide any desired combination of mixture enrichment and idling speed.

HS type carburettor

The recently introduced HS type is an improved version of the H type in which a flexible connection, a small diameter nylon tube, is used to carry petrol from the float chamber to the base of the jet tube (see 4 in Fig. 31). The jet slides inside a long tubular bearing and the chance of leakage at this point is much reduced in comparison with the H type.

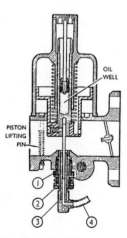

OIL WELL

PISTON LIFTING PIN

FIG. 31. The HS model is shown here in section

TUNING THE S.U. CARBURETTOR

Tuning is often inseparable from 'trouble-shooting' (as our American friends so vividly put it). The case-history of the particular car generally gives a clue to the particular defect. For example: the owner of the car complains that the car has never run right since he and his cousin Fred decarbonised it. He knows that Fred fiddled with the carburettor and he feels sure that an expert carburettor tuner will be able to put the matter right. The expert tuner immediately knows to check the compression pressures, since the valves may not be seating correctly, the head gasket may be blowing, or one or two valves may have no clearance at all. If this test is satisfactory he tests the induction manifold for leaks, in case the manifold gasket was damaged during replacement. Only when he is quite satisfied that the decarbonisation was well performed and that the ignition system is in a satisfactory state of tune does he turn his attentions to the carburettor.

At this stage one should examine the carburettor for likely defects and petrol leaks are usually the most obvious of these. The fitting of a new kit of seals to the jet assembly generally cures a drip from the bottom of the jet head; and the fitting of new fibre washers to the banjo connection of the petrol feed to the float chamber, or even tightening with a spanner, will stop leakage at this point. Flooding of the float chamber may be caused by dirt under the needle-valve, but a visible groove in the face of the needle valve can only be cured by fitting a new needle valve and seat. These are quite cheap and are fitted in a few seconds. A punctured float is another possible cause of flooding, but is rarely encountered nowadays. Shake the float while holding it near the ear and listen for the sound of petrol splashing around inside. A replacement is better than an attempted repair and is obtainable for a few shillings.

The carburettor piston requires cleaning at intervals of about 10,000 miles. The piston and suction chamber should be cleaned in petrol and carefully dried. The piston guide (the spindle at the top of the piston) should be smeared with light oil, but the remaining surfaces should be left dry. Care should be exercised in handling the piston to prevent damage to the needle, which is easily bent.

After re-assembly, a check should be made on the free unrestricted working of the piston. Any contact between the needle and the jet will prevent the carburettor from metering in a correct manner.

The piston should be lifted to the top of its stroke and allowed to drop. It should strike the bridge with a sharp click. Failure to fall freely usually indicates that the needle is rubbing on the side of the jet. If the needle is not bent, the fault lies in the centring of the jet. To re-centre the jet, first slacken the jet lock nut (5 in Fig. 28). The jet should now be free to move laterally to a small extent. The piston should be raised and dropped several times. It should now make a distinct click as it strikes the bridge. The lock nut should be re-tightened and the piston raised and dropped again to check that the position of the jet was not disturbed during the re-tightening. To centre the jet on the HD type carburettor, the float chamber and jet casing should be removed, the jet lock nut (21 in Fig. 30) slackened and the jet pushed to its highest position by hand. Dropping the needle into the jet should now centralise it and the lock nut should be carefully re-tightened.

When the carburettor is clean and working freely the damper reservoir should be filled with light oil (S.A.E. 20). Overfilling is no cause for concern, since surplus oil is pushed through the vent-hole in the filler cap. The carburettor is now ready for tuning and this operation in this narrow sense of the word, only means the setting of the idling mixture strength.

Vacuum gauge technique

The simple vacuum gauge offers us one of the easiest and most reliable methods of setting the slow-running mixture. Only one tapping is necessary in the manifold, even when entirely separate manifolds are provided on a multi-carburettor installation. If there is no existing tapping for the operation of windscreen washers or wipers, the wall of the induction pipe must be drilled and tapped to take a small-bore connection. Suitable 2 B.A. adaptors to take small-bore polythene tubing can usually be purchased from a Crypton or Redex tuning station.

The principle of the technique is as follows: at a given idling speed the vacuum 'pulled' by the engine against the nearly closed throttle-plate is a measure of the efficiency of combustion and the power given by the particular mixture supplied. When idling, all the power developed is used in overcoming internal friction and pumping losses. With efficient combustion of the correct mixture for maximum power, a given idling speed is achieved on the least amount of air, i.e. the highest vacuum in the induction system.

With a thoroughly warm engine, the throttle stop screw (or the slow-run valve on the HD carburettor) should be set to give a steady, fairly fast, idling speed—about 600-800 r.p.m. By weakening or enrichening the mixture, i.e. screwing the jet stop nut up or down (or turning the mixture screw, 8 on Fig. 30, anticlockwise or clockwise on the HD type carburettor) the highest possible reading on the vacuum gauge should be obtained. If the engine is in good condition, and the ignition and valve timing correct, a typical 'ordinary' modern engine will give a reading of 20-22 inches of mercury. A high-performance engine with a racing type of cam-shaft will give a much lower reading, since the valve timing pro-vided does not give efficient combustion at low speeds. Small changes in the engine speed during the operation can be ignored. If, however, the mixture was excessively weak or rich at the start, an adjustment of the throttle setting will become necessary to reduce the r.p.m. to the original value. The highest vacuum reading gives the maximum power setting at the particular speed. The profile of the needle, if correct, should maintain the correct mixture over the rest of the range. For an economical setting the jet stop nut should be screwed *up* about half a turn, i.e. three flats of the hexagonal nut (or half a turn of screw 8 on the HD type carburettor).

At the conclusion of the adjustments the adaptor should be re-moved from the manifold and a screw inserted to seal the hole.

The method recommended by the S.U. Company is to set the throttle to a fast idling speed and *adjust the mixture to give the fastest speed*. This method is particularly accurate when the car is fitted with a rev-counter or an electronic tachometer is available. Other-wise a small change in engine speed is difficult to observe.

When multiple carburettors are fitted

Before starting to tune a multi-carburettor engine the throttles should be synchronised. A short length of rubber tubing can be very useful as a stethoscope to listen to the intensity of the intake hiss at each carburettor mouth in turn. The air cleaners should first be removed and the clamping bolts on the flexible connections between the throttle spindles should be slackened to permit separate adjustment to the throttle settings of each carburettor. Equal intensity of intake hiss indicates equal air rates past the throttle. When this balance is achieved by adjustments to the throttle stop screws, the clamping bolts on the flexible connections should be re-tightened.

Special instruments for balancing multi-carburettor installations are now on the market. Rally Engineering of 7140 Seward Avenue, Niles 48, Illinois, make a good general purpose instrument to fit most common European carburettors.

For multi-carburettor installations any attempt in the early stages to approach the correct mixture by simultaneous adjustments of two or more carburettors can only lead to confusion. With three carburettors to set, both the *vacuum gauge technique* and the *mixture for maximum engine speed technique* become rather insensitive and the *piston-lifting test* is useful to check the accuracy of the mixture setting. If lifting the piston on one carburettor by about $\frac{1}{8}$ in. stops the engine or makes it misfire or falter and repeating the action on another carburettor has very little effect, the mixture on the first carburettor must be weaker than that on the second. As soon as the mixtures given by all carburettors appear to be alike, as indicated in this way, the effect of *simultaneous* adjustments of the mixture, two flats at a time on the jet stop nut, may be tried until the setting giving the highest vacuum reading is found.

Tuning after modification

Large changes in the idling mixture strength have very little effect on the mixture strength at full throttle, this being controlled almost entirely by the needle profile. Even at cruising speed a very rich setting of the idling adjustment does not produce much enrichment at these higher air flows. To produce enrichment or weakening over the whole range, or enrichment in one part of the range and weakening in another, a change to a needle of different profile becomes necessary. Even production engines differ, despite the close control made by modern foundries on cylinder head castings, and the S.U. firm usually provide three needle forms for each type of engine: a standard needle to cover more than 95 per cent of the engines and a weak and a rich needle for the odd-ones-out.

Major modifications to an engine in the form of supertuning often call for a radical change in needle form. A change to a different camshaft, induction or exhaust system, a large increase in compression ratio, a change in combustion chamber or port shape, even the fitting of larger inlet valves can upset carburation to such an extent that an entirely different needle profile is required. To illustrate how supertuning, in one aspect only—a change in cam form—can radically change the mixture requirement of an engine, we have reproduced below the needles recommended for the MGA and the M.G. Magnette.

TABLE 3

S.U. CARBURETTOR, NEEDLE PROFILES

Distance from needle shoulder, measured in eighths of an inch	Needle Designation		
	GS	4	EQ
	Standard needle for M.G.A.	Weak needle for M.G.A.	Standard needle for Magnette
0	0·089	0·089	0·089
1	0·085	0·085	0·085
2	0·0815	0·0814	0·0817
3	0·0785	0·0785	0·0785
4	0·0755	0·0761	0·076
5	0·0725	0·0737	0·0746
6	0·070	0·0714	0·0732
7	0·0675	0·0692	0·072
8	0·065	0·0668	0·071
9	0·0625	0·0645	0·070
10	0·060	0·062	0·069
11	0·0575	0·061	0·068
12	0·055	0·060	—

Note: The jet diameter is 0·090 in. in each case.

The first needle, type GS, is the standard one for the MGA: the second needle is the weak one for this engine. The third needle is the standard kind for the Z.B. Magnette. These two engines differ only in one respect (apart from a slight difference in the exhaust system). Basically the 'B' Series B.M.C. engine, they have the same compression ratio, the same size of valves, and both are fitted with twin H4 S.U. carburettors. The one important difference lies in the camshaft. That of the Magnette is designed to give good torque at medium speeds. The inlet valve opens 5° B.T.D.C. and

SCE G

closes 45° A.B.D.C.; exhaust opens 40° B.B.D.C. and closes 10° A.T.D.C. The valve lift is 0·322 in. The MGA camshaft is designed to improve volumetric efficiency at high r.p.m., with a small sacrifice in torque at medium and low speeds. Inlet opens 16° B.T.D.C. and closes 56° A.B.D.C., the exhaust valve opens 51° B.B.D.C. and closes 21° A.T.D.C. The valve lift in this case is 0·357 in.

The table shows what a large difference in needle form is required to compensate for the one major change in engine design. At Station 10 the change from standard to weak needle only requires an increase in diameter from 0·060 in. to 0·062, a decrease in effective jet area of 5·5 per cent. At this same station the standard Magnette needle has a diameter of 0·069 and if this needle were fitted to the MGA carburettor the reduction in jet area would be about 25 per cent and the mixture provided would be too weak for the engine to run.

If, then, we make any major changes by supertuning we are faced with the task of finding a new needle for the carburettor. The writer in his tuning work always uses a portable exhaust gas analyser which he can place in the passenger's seat and study the behaviour of the carburettor under all road conditions. Without this instrument the choosing of a new needle calls for patience and the expenditure of a few more shillings on needles. To assist the serious engine tuner the S.U. Company provide an invaluable publication, a list of all needles in production, giving their profiles in the manner of Table 3. With the aid of this, a stop-watch and several road tests the average tuner can soon find a suitable needle without any instrumentation to help him. This publication is AUC 9618 and can be obtained from the S.U. Carburettor Company for one shilling. Another useful publication, giving a full list of all cars using S.U. carburettors, with the sizes used, the needles used, and the types of piston spring fitted is List AUC 9631.

One final means of modifying the mixture characteristics of an S.U. carburettor still remains. A small change in the mixture strength provided over the whole range of air flows can be brought about by changing the strength of the piston spring. For identification purposes these are painted on the end coil in different colours to denote the spring strength. The normal range is given in Table 4 on p. 87.

TABLE 4

S.U. Carburettor, Piston Spring Strengths

Colour on end coil	Load, oz.
Blue	$2\frac{1}{2}$
Red	$4\frac{1}{2}$
Yellow	8
Green	12
Green and Black	5
(used only in H1 horizontal carburettor)	

A change to a lighter spring weakens the mixture. In hot climates or abnormally humid ones a metering change is sometimes required, and a change to a weaker spring will often suffice. Similarly at altitudes of 6000 ft. or more it is often necessary to fit a weaker spring to provide a correct mixture strength.

THE S.U. TWIN-CHOKE CARBURETTOR

This carburettor was developed for use on the Coventry-Climax FPF twin-cam engine and has proved to be an excellent power producer. It operates on the basic S.U. variable choke principle. An additional device, however, has been found necessary to adapt the principle to a one cylinder/one choke induction system. This addition is necessary because the full throttle impulses become too heavy when one choke per cylinder is used. As discussed earlier in this chapter these pulsations tend to draw a greater quantity of fuel from a given size of jet than the same air flow would accomplish if steady. The more vigorous the pulsations the more the fuel flow is enhanced. If then we tuned the carburettor to have the correct, i.e. slightly rich, mixture to suit full throttle conditions, the effect of a partially closed throttle plate would reduce the gas pulsations to such an extent that the part-throttle mixtures would be too weak. To compensate for this effect on the S.U. twin-choke carburettor a special device is provided which weakens out the mixture at full throttle, but is not effective at part-throttle conditions.

Fig. 33 shows diagrammatically how this new weakening device functions. The spring-loaded valve A is attached to the diaphragm B which acts as a dividing wall between the upper chamber C and the lower chamber K. Chamber C is maintained at approximately induction pipe vacuum by the connecting passage J and a similar passage (not shown) leading to downstream of the throttle-plate in the other choke. To reduce pressure pulsations to a minimum this chamber C is cross-connected with the corresponding chamber in the other carburettor. The lower chamber is connected by passage

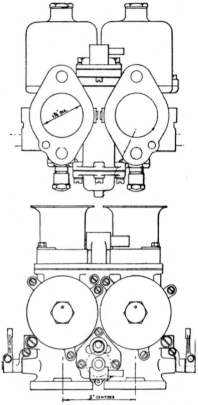

FIG. 32.   Side elevation and plan view
of the S.U. twin-choke carburettor.

L to the top of the float bowl, and the volume H below the spring-loaded valve is connected via jet G to the zone of constant depression (approx. 8 in. of water) that always exists on an S.U. carburettor in the space between the carburettor piston and the throttle-plate.

When the engine is running at part-throttle conditions the de-

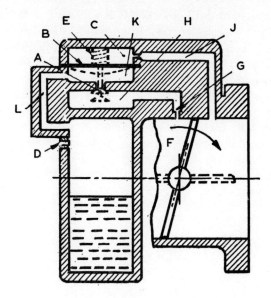

FIG. 33. Weakening device, as used on the twin-choke S.U.

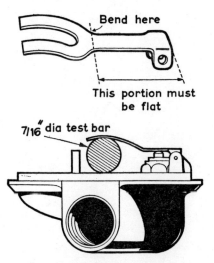

FIG. 34. How to check the float lever setting for correct fuel level

pression transmitted to chamber C is sufficient to force valve A closed. In this condition the pressure above the fuel in the float chamber is sensibly atmospheric. Under full-throttle operation the depression in chamber C has fallen to such a value that the spring E opens the valve and permits an air bleed to exist from atmosphere into the constant depression zone F. This flow of air is from the air-bleed D, through passage L, chamber K, volume H and jet G. The sizes of the two drillings D and G are designed to produce the desired lowering of pressure on the fuel in the float chamber to weaken off the mixture at full throttle by an amount that adequately compensates for the over-enrichment caused by pressure pulsations.

CHAPTER SIX

Fixed Choke Carburettors

WORKING PRINCIPLES OF FIXED CHOKE CARBURETTORS

AN OLD American song tells us, 'There ain't no law 'gainst sucking soda thru' a straw.' In this chapter the writer is more concerned with the *natural laws* governing the passage of liquids through jets and he has been struck by the similarity between the above method of transferring liquids to the human stomach and the working of the simple fixed-choke carburettor. When Lesley, in Fig. 35,

FIG. 35

applies a suction to the upper end of the straw (by the expansion of her lungs) atmospheric pressure forces liquid up the straw into her mouth. The greater the depression applied to the straw the greater will be the flow of liquid into her mouth. In the simple carburettor (see Fig. 36) the straw is replaced by the main jet and its spray tube (the use of a removable jet provides us with a ready

91

means of changing the mixture characteristics, otherwise a simple small-bore tube would suffice). The lungs are replaced by the pistons and the depression created by the descending piston is intensified by means of the venturi tube, or choke, in the carburettor.

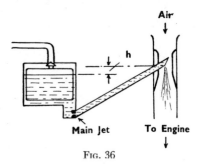

Fig. 36

The greater the gas velocity through the choke, the greater will be the flow of fuel along the spray tube.

Unfortunately the flow of fuel increases in proportion to the *velocity* of the air flowing through the choke. Since the density of the air in the choke falls as the velocity increases, the mass-flow of air does not increase at as high a rate as the flow of fuel. Thus if we choose the sizes of the main jet and the choke tube to give a mixture of 15 lb. of air to 1 lb. of fuel when the throttle is only partially open we find that if we open the throttle until the mass air rate is doubled the fuel rate is increased about $2\frac{1}{2}$ times. The mixture is thus enriched to give 1 lb. of fuel to every 12 lb. of air. At four times the air rate the mixture would become excessively rich. The extent of this enrichment with increased air rate can be varied slightly by changes in jet shape or by increasing or decreasing the level of the fuel in the float chamber relative to the tip of the spray tube. These variations, however, are never large and cannot influence the general rise in the mixture strength curve.

The enriching effect is exactly the opposite to our requirements, since our engine needs a rich mixture when idling and a weak one when cruising.

Mixture compensation

The process of correcting this enriching tendency of the simple-jet carburettor is usually called mixture compensation and the method evolved by early carburettor designers involved the use of a device called 'the submerged jet'. To understand the principle of the submerged jet it is helpful to consider again the case of Lesley and her straw. Left to her own devices Lesley knows from experience that the end of the straw must be completely immersed to get a full flow of liquid. To her simple straight-thinking little mind this would be called a 'submerged jet'. The perverse technician, however, uses this term for a jet which, to our way of thinking is not submerged. If, for instance, the glass is almost empty and the lower end of the straw becomes uncovered, we are presented, for a brief moment, with the conditions of the submerged jet. The condition is only brief, since the liquid soon disappears and we are left with only air passing up the straw. If we wish to prolong the submerged jet condition, it is necessary to trickle a supply of fresh liquid down the side of the glass, at such a low rate that the end of the straw does not become covered (see Fig. 37). The noise made by

FIG. 37

sucking a mixture of air and liquid is condemned by all the best books on etiquette. It is made, nevertheless, by some of the best carburettors.

The principle of the submerged jet (the compensating jet of the Zenith carburettor) is shown diagrammatically in Fig. 38. Lesley's glass becomes the well and the compensating jet provides the trickle of liquid to prevent the spray tube (the straw) from running dry. This compensating jet, used alone, would feed a mixture that would gradually weaken as the air rate increased. The compensating jet is not subjected to choke depression, the depression being 'broken' by the well which is open to atmospheric pressure. What-

ever the choke depression the fuel feed from the compensating jet
is constant. The amount of air drawn from the well, however,
increases with choke depression and the overall effect is a gradual
weakening of the mixture as the choke depression increases. Since
this is the opposite effect to that given by the simple main jet, it
would be reasonable to expect a combination of the two types of
jet to provide a fairly constant mixture at all choke velocities. How

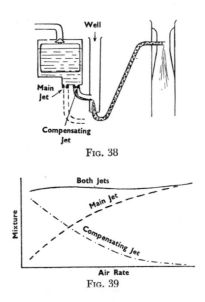

Fig. 38

Fig. 39

this works is shown graphically in Fig. 39. This system of compensation has been used on various types of Zenith carburettor and
is still used on some of the current models.

On the Solex, Weber and Stromberg carburettors, and on certain
Zenith types based on the Stromberg principle, only one major jet
is used. This jet system resembles the submerged jet principle, with
the difference that the air entering the well is restricted by an air
correction jet. There is also a second variable restriction called an
'emulsion tube'.

To understand the Solex 'air-bleed' principle we must approach
it in easy stages. Let us as a first step consider what happens if we
restrict the admission of air to the well of a submerged jet, as shown
in Fig. 40. If we were to close the top of the well completely the

well could fill up with fuel, the jet would be submerged to full choke depression and the flow behaviour would be that of the simple main jet, i.e. the mixture would become richer as the air rate increased. If, however, we allow air to bleed into the well through a relatively small restriction, as in Fig. 40, there will be a certain

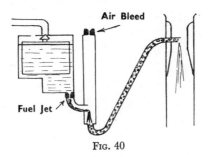

FIG. 40

depression in the well—not the full choke depression but a somewhat lower value, depending upon the size of the air-jet. The flow from the fuel jet with this system will be less than with a simple jet, but greater than with a submerged jet. At first glance it would appear that we have in this device, if only we choose our air and fuel jets correctly, a method of maintaining a fairly constant mixture strength over the whole cruising range. Unfortunately we are foiled again by the metering habits of the air jet, which is, in effect, a crude form of venturi. On a *volume* basis the mixture remains fairly constant, but on a weight basis it becomes richer as the air rate increases. To correct this tendency, it is necessary to make the air bleed increase in size as the air rate increases. This is the main purpose of the emulsion tube, shown in Fig. 41. In this tube holes are drilled at different levels. At low air flows, the difference in pressure between the inside of the tube and the outside is small and air is admitted only through the upper ring of holes. As the air rate through the choke increases, the depression in the choke, relative to the smaller depression inside the emulsion tube, begins to rise. Thus the level of the fuel inside the emulsion tube falls and uncovers the second ring of holes. In this way the air metering system is variable and by choosing suitable sizes of emulsion holes and air correction jet we can provide sufficient correction to give a sensibly constant mixture strength over the desired range. Fig. 41 serves to

illustrate the Solex principle of correction but, in the majority of
Solex models, the jet and emulsion tube assembly are situated in the
centre of the choke, with the air correction jet in the high-pressure
region at the inlet to the choke and the feed tube in the point of

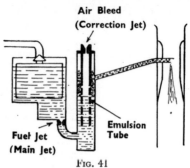

Air Bleed
(Correction Jet)

**Emulsion
Tube**

**Fuel Jet
(Main Jet)**

FIG. 41

maximum depression. In the Weber and Stromberg carburettors it
is at the side, in the manner of Fig. 41.

Many carburettor books stress the importance of the emulsion
tube as a device for breaking down the fuel into small droplets to
assist vaporisation in the induction tract. Its function as a variable
air metering device is, however, far more important.

Starting

When starting an engine fitted with a fixed-choke carburettor
the air flow through the choke is so low that unless special means
are taken to increase it the choke depression will not induce sufficient
fuel to flow through the jets for our purpose. One of the earliest
starting methods was to hold the hand over the carburettor air
intake and crank the engine at the same time. The slipped disc
had not been invented in those days, but the occasional backfire
served to demonstrate that this was not the ideal method of starting
an engine. This crude method was soon replaced by a mechanically-
operated flap at the intake. This is—or was—commonly called a
strangler in this country and a choke in America. With the carburettor mouth almost completely blocked in this way, the full
engine depression extends past the throttle plate as far as the intake
itself. The extent of the depression given by a strangler will depend
upon the cranking speed, but in general it will be between 1 and
3 in. of mercury. This depression is sufficient to induce a flow of

fuel from the main and compensating jets. The mass flow of air being relatively low the resulting mixture will be very rich. This method is used on certain Zenith models and, in conjunction with an automatic control, on both American and British Stromberg designs.

A modern method, which seems to offer a better control over the degree of richness than the strangler, is the provision of a starter carburettor. This is a simple auxiliary carburettor with an air jet and a fuel jet feeding into a passage leading to the engine side of the throttle plate. The Solex Bi-starter is a good example of this method.

Idling

To provide a rich mixture for idling, all fixed-choke carburettors rely on what is called 'throttle-edge control'. With the throttle adjusted for slow running a small flow of air can pass round the edge of the throttle as shown in Fig. 42. The engine depression is

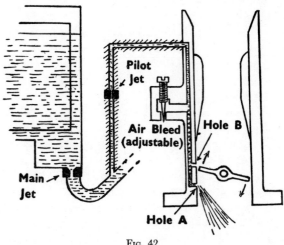

FIG. 42

high and a flow of fuel, metered by the pilot jet, is drawn through the small hole A into the stream of air at the throttle edge. Air will also enter the passage from the pilot jet through the upper hole B, since this is above the throttle and communicating with a region of high pressure. Variation of the mixture strength can be arranged in two ways. Sometimes the size of the air bleed can be regulated by a tapered screw, as in Fig. 42. In other types of carburettor a tapered needle enters the passage taking the mixture to hole A. As the throttle is opened the depression begins to fall and the flow of fuel from hole A is reduced. The upper hole B is now subjected to the full depression downstream of the throttle edge (following arrow) and instead of acting as an air bleed to A, now becomes a second fuel jet, augmenting the diminishing flow from A. In this way the second hole B serves to bridge the gap between the fading out of the pilot jet and the coming into action of the main or compensating jets. Careful design is necessary to avoid a flat spot as the throttle is gradually opened from the idling position.

Full-throttle operation

For full-throttle operation we require a maximum power mixture of between 12 and 13 to 1. This is a much richer mixture than we would use for part-throttle running, but it is necessary if we wish to obtain the full power the engine can provide. The danger of valve-burning is also greater at the higher temperatures of full-throttle operation and it is advisable to keep the mixture on the rich side of the chemically correct value if valve oxidation is to be kept to a minimum. Where distribution errors are thought to be bad, as in, say a six-cylinder engine fed by a single carburettor, a rich mixture is essential at full-throttle.

There are many ways in which the mixture can be enriched at full throttle. The older types of Zenith carburettor relied upon the correct choice of main and compensating jets to give a richer mixture at high air rates. With such a carburettor, however, one cannot provide an economical mixture at, say, quarter-throttle and 4000 r.p.m. and a rich mixture at full-throttle and 2000 r.p.m., since the air rates could be identical in the two cases.

There are two methods in general use on modern carburettors for providing the richer mixture for full-throttle operation. In the first method an economy valve, operated by engine depression, admits extra air to the capacity well of the submerged jet during part-throttle running. As soon as the throttle is wide open the falling depression causes the diaphragm of the economy device to close, cutting out the extra air and thus enriching the mixture. This method is used on VN and VIG type Zenith carburettors. In the second method the jets are chosen to give economical mixtures over the required part-throttle range and the enrichment for full-throttle is provided by what is commonly called a 'power jet'. This jet may be brought into action by a depression-controlled diaphragm (Stromberg), by a valve lifted from its seat at a certain choke air speed (Weber) or by a mechanical connection to the throttle itself (Zenith and Solex).

Acceleration

The capacity well of the submerged jet system is sometimes called the acceleration well. On earlier designs of Zenith carburettor this well served as a reserve of fuel to provide a richer mixture during acceleration. When the engine is idling the fuel drawn from the main and compensating jets, is a negligible amount. Consequently the capacity well can fill up with fuel almost to the level of that in the float chamber. If the throttle is suddenly opened the increased depression in the choke quickly empties the well, but until this happens the compensating jet does not behave as a true submerged jet, but gives an enhanced feed-rate, thus helping to give the required enrichment during the critical second after the snap-opening of the throttle.

The acceleration of the modern car, especially the modern sports car, is now so fierce that a more rapid and almost instantaneous injection of extra fuel is necessary if a flat-spot is to be avoided. Modern fixed-choke carburettors are sometimes provided with what is called an acceleration pump. In one example this takes the form of a plunger operating inside a cylinder. During normal steady running the plunger is retracted and the cylinder fills up with fuel drawn through a non-return valve from the main jet supply. Movement of the plunger injects the contents of the cylinder into the choke during acceleration. This type of pump is operated by a mechanical linkage from the throttle lever. In another design the pump is a diaphragm which is pulled back against a spring by the action of the depression upstream of the throttle. In this position the capacity chamber of the pump is filled with fuel. Sudden opening of the throttle destroys the depression in the induction tract, which in turn releases the depression which is holding the diaphragm against the force of the spring. This spring moves the diaphragm against the fuel stored in the chamber and injects it at a high rate into the choke. In other modern carburettors the diaphragm is mechanically connected to the throttle lever, an adjustment of the length of stroke being incorporated to permit a reduced amount of fuel to be injected during summer motoring.

ZENITH CARBURETTORS

A few years ago the Zenith Carburetter Company introduced a new range of Zenith-Stromberg carburettors operating on Stromberg principles. Before this time there was only one basic design—one 'true Zenith' carburettor. This instrument metered fuel over most of the working range from two jets, a main and a compensating jet. We have already shown how the addition of a submerged jet of the correct size can provide the necessary degree of compensation to give a resultant mixture that is sensibly constant over the required range of air rates. Zenith carburettors of this type are still in general use and it is proposed now to describe the Series VN carburettor as a typical example of the true Zenith pattern.

The V-type Zenith in its many variants has been in production since 1933. It has been a most popular carburettor and the excellent economy of numerous Austin and Vauxhall cars owes much to the sound basic design of the Zenith carburettor fitted. The current fashion of the falling bonnet-line places a restriction on carburettor height and the newer VN instrument is only two-thirds the height of the older Series VIG carburettor.

Starting and idling

The spring-loaded strangler flap (14 in Fig. 43) is released by operation of the dashboard choke control. This closes the carburettor air intake and, by applying full induction depression to the idling jet system, provides a rich mixture for starting. An interconnecting rod between strangler lever and throttle spindle opens the throttle by a regulated amount to give a fast idle. Increase of engine speed produces an increased depression. This, acting on the unbalanced strangler flap, opens it to a greater extent thus reducing the degree of enrichment. The extent of the opening is controlled by the strength of the strangler spring. This type of strangler is usually called an automatic one.

The vehicle can be driven away on full-throttle if desired, but

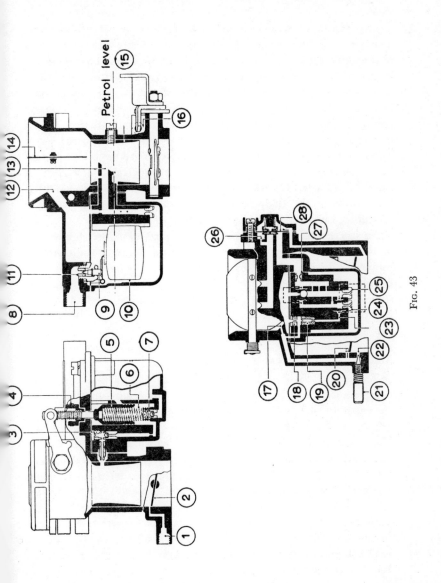

Petrol level

FIG. 43

the strangler control should be returned to normal as soon as possible.

The Zenith Company never seems to have made up its mind whether to control the idling mixture strength by a tapered screw in the idling orifice (see 21 and 22 in Fig. 43) or by a tapered screw in the air bleed. Even in the same series of carburettor—the VIG, for example—both types of idling adjustment can be found. The first method, as used on the Series VN, is usually called volume control. The second method is described as air regulation. On a downdraught Zenith identification is easy. If the adjusting screw is nearer the top than the bottom of the carburettor body, it is an air regulating screw and a richer mixture is given if the screw is turned clockwise. On the other hand a regulating screw mounted low down on the body is a volume control screw and a *weaker* mixture is given by clockwise rotation.

An interesting feature of the VN carburettor is the duplication of the so-called progression orifices in the idling circuit (called by-pass orifice on the Solex carburettor). A cynical carburettor expert once remarked to the writer, 'If in doubt, drill another hole somewhere.' He modestly omitted to mention that he had twenty-five years of carburettor work behind him to tell him *where* to drill this hole!

Main metering system

The traditional Zenith metering system using a main jet and a compensating jet was described earlier. In older types of Zenith the capacity well is open to atmospheric pressure. In the Series VN admission of air to the well is controlled by two air bleeds in series. The main air bleed, 27 in Fig. 43, is also sometimes described as 'ventilation screw to capacity well'. The effect of this restriction is greater at high air rates and a change to a smaller main air bleed will enrich the mixture, but largely at the 'top end'. On a standard engine one never need change the size of this bleed. Between this bleed and the atmosphere is placed a second restriction which is under the control of the economy device. Under cruising conditions with a high vacuum in the induction manifold, the diaphragm 28 is pulled to the right and a free unrestricted passage admits air to air bleed 27. At full throttle and a relatively low manifold depression the diaphragm is pushed to the left under the action of its spring and the large-diameter passage is closed. The only connection between the main air bleed 27 and atmosphere is now through the secondary bleed 26. The depression in the capacity well is thus greater at full throttle and the mixture supplied to the engine is correspondingly enriched. In keeping with this, the sizes of main and compensating jets are chosen to give a weak cruising mixture.

A simple piston-type acceleration pump is used, with non-return ball valves at inlet and outlet. The speed of discharge is controlled by the size of the pump jet, the amount discharged by the length of stroke. A simple means of altering the pump stroke is provided by the use of a two-position stop. This takes the form of a small cast block (4 in Fig. 43) which can be turned through 180° from the position shown in Fig. 43. The stroke of the piston is limited by the lower edge of the arm striking the inner edge of the block. In Fig. 43 the block is in the long-stroke position.

Tuning the Series VN carburettor

Setting the idling mixture and idling speed of the VN instrument, by means of the volume control screw (21 in Fig. 43) and the throttle stop screw (16 in Fig. 43) needs no elaboration. Any of the recognised methods can, in fact, be employed. Where multi-carburettors are fitted the usual care should be taken to ensure that the throttles all open simultaneously, an audible check for equal hiss being made by means of a short piece of rubber tubing. Before the commencement of tuning, the volume control screws on all carburettors should be set so that each is open by the same amount. From this point all adjustments of mixture strength should be made by equal amounts to each screw.

In cases of special tuning, such as the fitting of a non-standard carburettor, one sometimes finds that a change in metering is required to give satisfactory idling. On the Series VN the pilot or idle jet (19 in Fig. 43) can be changed as a first step. A change to a larger or smaller air bleed, 18, may also be necessary to match the upward or downward change in idle jet. On this particular V-type carburettor all jets are carried in a detachable emulsion block. Therefore to reach the idle jet the four screws in the float chamber cover should be withdrawn and the float chamber removed horizontally to release the beak of the emulsion block from the main carburettor casting. When the fixed air bleed 17 is adequate for the particular application a blank plug is fitted above the idle jet, indicated in Fig. 43. Removal of this plug (or air bleed) gives access to the idle jet.

The main and compensating jets are screwed into the base of the emulsion block. To remove this block, the float and float arm should be removed and the screws holding the emulsion block to the float chamber extracted. It is important to note when re-assembling that the float and arm are both marked with the word 'top'. Changes to these jets are usually only necessary after major tuning modifications and it is imperative to check that the correct type of jet for the VN carburettor is supplied. *Both main and compensating jets should be cadmium-plated.* Plain brass jets, similar in external dimensions, but with different flow characteristics, are occasionally fitted in error.

It will be remembered that the characteristic behaviour of a main jet used alone is to provide a richer mixture with increase in air rate. The flow from a submerged jet, i.e. one feeding a well open to atmosphere, is constant. Such a jet, therefore, as a separate feed to a choke would supply a mixture that became weaker with increase in air rate. A suitable combination of the two types of jet can provide a fairly constant mixture over a wide range of air rates. This is the Zenith system of compensation, shown in the dotted square (24 and 25) in fig. 43. It will be seen from this that a change in size of main jet will have a more marked effect on the overall mixture strength at high air rates than at low. The compensating jet on the other hand has a greater influence at low air rates. Thus if our road tests indicate a lack of power at the top end a larger main jet should be fitted. If, for example, we find that the best top-end performance is given by an increase in main jet of 10, then the next step is to note the effect of a decrease in compensating jet. The method of tuning a new installation is largely the same as for the Solex carburettor, with the important difference that the Solex main jet has the greatest effect at low air rates and the Zenith at high. As with so many things, experience reduces labour and the experienced tuner arrives at the correct jet settings by far fewer steps than the beginner.

The use in the Series VN carburettor of an air bleed to ventilate the capacity well gives us another means of varying the mixture strength. A reduction in size of this air bleed (27 in Fig. 43) increases the depression applied to both main and compensating jets, increases the flow and enriches the mixture. The effect is greater at the top end of the range (as with the Solex air correction jet). It is seldom, however, that a change is required in the size of this air bleed.

The accelerating pump has two possible variables. The rate at which petrol is injected into the choke is controlled by the size of the jet. The smallest jet that can be used without hesitation on snap accelerations is the correct one. It is not, of course, normally changed on standard installations. The second variable is the pump stroke. The movable block has only two positions: a short stroke, which is the summer position, and a long stroke, or winter position.

Effect of altitude

In Europe the effect of altitude changes on carburation is quite small. Even in the Alps, although power is lost on the higher passes, carburation is not seriously affected. In some parts of the world, however, a car may spend most of its working life on a high-altitude plateau and a change from the standard jet settings is then advisable. For an altitude of 5000 to 7000 ft. with a medium-size carburettor, one would make a reduction in both main and compensating jets of half a size. A Zenith full-size change of jet is taken as five numbers. Half sizes are stamped on the jets to the nearest whole number. A change from 60 to 55 would be a full-size change; from 60 to 57 a half-size. To maintain sea-level mixtures at an altitude of 7000 to 10000 ft. a full-size reduction in both jets would be required.

Series W carburettor

In recent years the business association between those two pioneer companies, Zenith of Great Britain and Stromberg of America, has led to the introduction of the W series of Zenith carburettor in which Zenith economy has been incorporated into what is essentially a Stromberg instrument. We have chosen the WIA as a suitable example of this new series.

Since the starting and idling arrangements on this carburettor differ only in minor details from those on the Series VN no purpose is served by describing them. In working principle the main metering system of the WIA resembles the Solex carburettor, with which some readers will be familiar and which will be dealt with later. The main fuel supply is metered through the main jet, 16 in Fig. 45. The fuel from this jet enters a capacity well containing an inclined tube. This tube, 19 in Fig. 45, is called the main discharge jet and is provided with a row of emulsion holes on the lower side. Air is metered to the well from an acorn-shaped air-bleed known

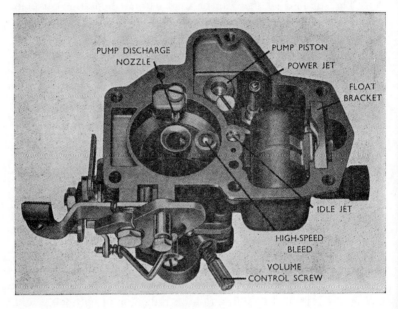

FIG. 44

as 'the high-speed bleed' (see Fig. 44). (This corresponds to the Solex air correction jet.) This main metering system, in fact, functions in the same way as the Solex system. As the choke velocity increases and the choke depression rises the level of the fuel in the well falls and the number of holes exposed in the main discharge jet increases. In this way increased air dilution gives a measure of air correction to the enriching tendency of the main jet.

A good Stromberg feature retained in the Series W carburettor is the use of a secondary choke tube or venturi (20 in Fig. 45). The double choke arrangement is acknowledged to give improved atomisation and to improve the torque of the engine at low speeds.

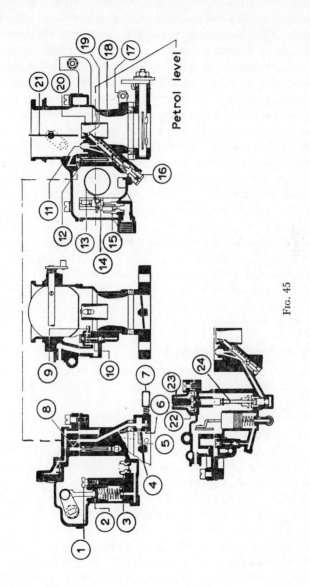

Petrol level

Fig. 45

The Series W carburettor is designed so that economy settings can be used for the main metering system, the extra fuel for full-throttle operation being provided by a power jet. On the WI model this jet is opened by mechanical means at a point near full-throttle. On the WIA, as shown in Fig. 45, the power jet is opened by a plunger valve attached by a long rod, 24, to the flexible diaphragm 22, which is mounted in the float-chamber cover. Under part-throttle conditions the induction vacuum acting on the upper face of the diaphragm is high enough to keep the valve on its seat. At full throttle the depression is not sufficient to hold the valve closed and it is forced open under the action of control spring 23. This admits fuel from the float chamber to the well through the jet at the bottom of the valve chamber.

A piston-type accelerating pump is fitted. This discharges through a non-return valve, 10, into a nozzle, 9, situated above the main choke tube. The pump is connected mechanically to the throttle spindle, one of the connecting links being provided with three holes to give a choice between a short, medium or long stroke to suit the prevailing atmospheric temperatures. During full-throttle acceleration extra fuel is supplied from two sources, the accelerating pump and the power jet.

Tuning the Series W carburettor

When the carburettor has been fitted as original equipment, tuning is normally confined to setting the idling mixture in the normal way by means of the volume control screw (see Fig. 44) and setting the idling speed with the throttle stop screw.

When economy is sought a half-size smaller main jet can be fitted and the effect on slow-speed behaviour noted. Occasionally a slightly smaller power jet can be used without any noticeable drop in high-speed performance. This change is ill-advised, however, in the case of a hard-driven rally car. Changes in these two jets, together with the choice of stroke for the accelerating pump, are the only normal carburettor variations necessary for tuning a standard engine. Access to the main jet is given by removing the hexagon plug in the float chamber base. To reach the idle tube (pilot jet), accelerating pump and power jet it is necessary to extract the six screws holding the float chamber cover. Unlike the Solex air correction jet the high-speed bleed is not detachable and the only air-flow variable is the main discharge jet. These jets are available in a range of bore diameters and air-bleed hole diameters. The discharge end can also be obtained open or masked. The Series W carburettor is not an easy model for the amateur to tune to a new installation and the advice of the carburettor maker should be sought before attempting this work. When tuning for maximum power on a new installation or a modified engine the maximum choke tube consistent with reasonable low-speed torque should be fitted. The small venturi cannot be changed.

No mechanical device is perfect and the W-type Zenith is no exception. One improvement on the older Stromberg carburettor is in the detail design of the float mechanism. In the older model the needle valve was free to spin on its seat, resulting occasionally in rapid wear. In the new design the needle valve is hinged to the lever to prevent rotation. All piston-type accelerating pumps can be put out of action by dirt in the petrol. The symptoms are hesitation during acceleration and the remedy is obvious. Poor acceleration can also be caused by a blocked power jet. Failure of the diaphragm would give similar symptoms. Excessive fuel consumption can also be caused by dirt trapped under the plunger valve, since this will permit leakage around the greater part of the valve face at all times.

Faulty behaviour of the carburettor can sometimes be traced to incorrect assembly. The W-type carburettor is far from foolproof in this respect. Replacement of the float chamber cover calls for special care. The pump lever in the cover should be carefully aligned with the top of the pump piston and the power jet operating plunger in the cover should be centrally located above the spring-loaded valve spindle, 24 in Fig. 45. Occasionally excessive fuel consumption can be traced to the disappearance of the small ball at the base of the pump discharge valve, 10 in Fig. 45. This ball is easily lost, especially if the body of the carburettor is inverted when trying to extract the holding screw (see Fig. 44). Downward movement of the pump piston after the discharge valve has been removed results in a ballistic ejection of the ball—amusing only to the bystander. The loss of the ball leaves an open channel to the pump chamber and, above a certain depression in the choke, the non-return valve 3 is opened and petrol is injected straight into the choke.

Mixture correction

To correct the mixture for altitude changes the reduction in size of the main jet should be approximately the same as given for the Series VN carburettor. The only other change necessary is a decrease of half a size in power jet for an altitude of 7000 to 10,000 ft.

The service department of the Zenith Carburetter Company at Honeypot Lane, Stanmore, Middlesex, is always on hand to deal with tuning queries. Full details should be sent, giving type number and model of car and the reference letter and numbers stamped on the carburettor.

SOLEX CARBURETTORS

For many years Solex Ltd have concentrated almost exclusively on downdraught carburettors, the type fitted to so many modern cars. The Solex main metering system, using a single (main) petrol jet and an air correction jet was described in detail earlier in the chapter. The new types that appeared after the war revealed a new awareness of the requirements of the export market. Much attention was given to the problem of dust exclusion. The float bowl was vented internally by a ducting opening into the air intake. With the exception of the air to the starter carburettor, all air entering the engine passed through the air filter. In a very dusty climate this is essential to prevent rapid engine wear. With an external vent to the float bowl, the gradual choking of the air cleaner, before it is cleaned or replaced, leads to a gradual enrichment of the mixture. With the internally vented design, the depression induced by a choked air cleaner is also transmitted to the fuel in the float bowl and the undesirable enrichment is avoided. Several of the latest designs are fully dust-proofed, air to the starter carburettor being admitted by means of a duct cast into the inside face of the carburettor inlet flange.

Several basic Solex types will now be described, starting with a simple pattern and working up to something more complex.

Type 32 BI

As in all Solex carburettors, the '32' in the type designation indicates the 'bore' of the carburettor, i.e. the outside diameter of the choke tube. Fig. 46 is not a true cross-section, but is excellent for our purpose. The main jet, G, is mounted in a hexagon-headed plug, *t*. The emulsion tube, *et*, is held in place by the correction jet *a*. The pilot jet, *g*, draws its petrol supply from the reserve well (or acceleration well), *v*, and the idling mixture from the pilot jet and the pilot jet air bleed, *u*, passes down the vertical duct to the idling mixture orifice *io*. The relative sizes of *g* and *u* control the strength of the idling mixture. The quantity is controlled by the

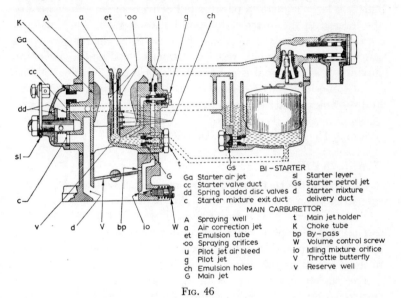

BI - STARTER	
Ga Starter air jet	sl Starter lever
cc Starter valve duct	Gs Starter petrol jet
dd Spring loaded disc valves	d Starter mixture
c Starter mixture exit duct	delivery duct

MAIN CARBURETTOR	
A Spraying well	t Main jet holder
a Air correction jet	K Choke tube
et Emulsion tube	bp By–pass
·oo Spraying orifices	W Volume control screw
u Pilot jet air bleed	io Idling mixture orifice
g Pilot jet	V Throttle butterfly
ch Emulsion holes	v Reserve well
G Main jet	

Fig. 46

position of the volume control screw, *W*. The manner in which the by-pass orifice, *bp*, helps to bridge the gap before the main jet spraying system begins to function has already been described.

The starter carburettor on the BI type is modern in that it draws its air supply internally, but is identical in principle with the Bi-starter design introduced before the war. The lever, *sl*, is connected by a cable of the push-pull type to the 'choke' knob inside the car. As the name implies, the Bi-starter has two positions, a fully extended position giving a mixture strength of about 4 to 1 for cold starting and an intermediate position giving about 9 to 1 for cold running. Petrol is metered by a starter petrol jet, *Gs*, and a starter air jet, *Ga*. *Gs* is conveniently placed in the side of the float bowl. This is the jet to change when it is found necessary to provide a richer or a weaker starting mixture. The correct sizes of the starter jets are found by extensive tests at the maker's factory and it is not often that one needs to change them. The petrol metered by *Gs* passes by internal ducting through the hole, *cc*, in the inner metering plate, *dd*, where it mixes with the air from *Ga*. The mixture passes back through the metering plate by hole *c*, and passes down the delivery duct, *d*. In the cold starting position the hole *cc* is large enough to pass the full delivery of *Gs*. Movement of the lever *sl* to the intermediate position rotates the inner plate *dd* until a smaller hole is opposite the duct from the starter petrol jet, *Gs*. This smaller hole restricts the petrol feed and weakens the mixture to a value that is more suitable for a cold running engine.

A two-position starter carburettor is obviously not an ideal device and the Solex starter carburettor has appeared in several improved variants in more recent years. A recent design as fitted to the Triumph Herald carburettor is described below.

Type B 28 ZIC-2

By moving the correction jet and emulsion tube assembly to the side, as shown in Fig. 47, the effective cross-section for a given choke size is increased. The reduced surface presented by the spraying assembly also helps to improve volumetric efficiency. The main spraying circuit and the idling circuit are almost identical to those in the Type BI carburettor.

The starter carburettor is of the dustproof zero progressive starter pattern and is designed for operation down to 0°F. (Since coming to live in America the writer is no longer impressed by zero. Many times he has had to make starts from temperatures of 10 or 15 degrees below zero Fahrenheit. The local product does this remarkably well.) When in the full-rich position the only air supplied to the engine, apart from leakage past the throttle-plate, is metered by a small air bleed *Sb*. This air, admixed with petrol from the starter jet, *Gs*, passes through the channel, *D*, across a slot in the starter disc valve, *Dd*, into duct *d*. The starting mixture strength can be varied by a change in the value of *Gs* and in very cold climates a *Gs* setting to give a 1 to 1 air/fuel ratio is used.

As soon as the engine fires the increased induction depression pulls the air valve, *C*, off its seat against the action of spring *X* and additional air is admitted through the drilling, *Ga*, in the disc valve, *Dd*. This drilling is in effect the starter air jet. As the engine warms up a gradual reduction in mixture strength can be achieved by rotating the disc valve, *Dd*, since this controls the amount of mixture escaping from channel *D*.

An added refinement in the starter carburettor is the provision of an enrichment device which functions right up to full-throttle as long as the choke control is pulled out. This consists of a special

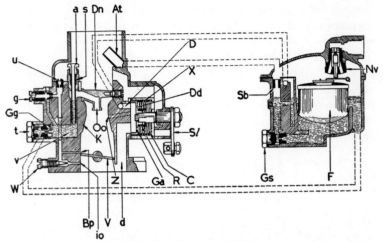

FIG. 47. Model B 28 Zic-2

Key to Diagram: At, Float chamber vent; a, Correction jet; Bp, By-pass orifice; C, Starter air valve; D, Starter petrol channel; Dd, Starter valve; Dn, Discharge nozzle; d, Starter outlet channel; F, Float; Ga, Starter air jet; Gg, Main jet; Gs, Starter petrol jet; g, Pilot jet; io, Idling orifice; K, Choke tube; Nv, Needle valve; Oo, Spraying orifice; R, Starter valve spring; Sb, Starter air bleed; S*l*, Starter lever; s, Emulsion tube; t, Main jet holder; u, Pilot jet air bleed; V, Throttle; v, Reserve well; W, Volume control screw; X, Starter air valve spring; Z, Quick drive-away channel.

drilling between the choke tube and the starter carburettor (see Z in Fig. 47). With the throttle open and a high enough air rate to induce a substantial depression at the throat of the choke tube petrol will be drawn from the slot in the disc valve, *Dd*, and injected into the choke tube. The provision of this additional fuel permits a car to be driven away safely on full-throttle with a cold engine. When idling (and with the starter carburettor in action) the drilling, Z, serves as an additional air bleed into the starter carburettor. Complete closure of the dashboard control rotates the

disc valve, *Dd*, until the fuel and air supplies to the starter carburettor are completely cut off.

Type C. 35 APAI-G

This dual-throat two-stage carburettor is an effective answer to
an old problem. If we fit a relatively large choke tube to try to
extract a little more power the low speed torque inevitably suffers.
If this is taken to extremes we can finish up with an engine with a
very narrow useful torque range and a most unattractive road vehicle,
especially on to-day's crowded roads. This Solex type is one of the
options on the Alfa Romeo Giulietta Spyder. Unlike the twin-
choke carburettors used on racing engines, the two choke tubes are
arranged to feed into a common manifold. At low air rates only
the primary choke is in operation, while at high air rates both
chokes are in use. In this way good atomisation is given over a
very wide operating range. The method is widely used in America.

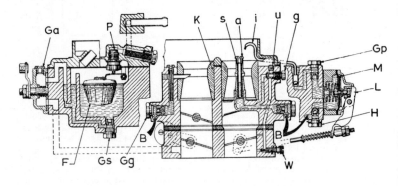

a	Correction jet	i	Pump injector
B	Fuel inlet	K	Choke tube
F	Float	L	Pump lever
g	Pilot jet	M	Pump membrane
Gg	Main jet	P	Needle valve
Gp	Pump jet	s	Emulsion tube
Gs	Starter petrol jet	u	Idling air bleed
H	Pump valve	W	Volume control screw

FIG. 48

In Fig. 48 the primary choke is on the right and is seen to have
all the normal jets and circuits of a single-choke Solex carburettor.
The device on the extreme right is the accelerating pump which

will be described later. The primary throttle, on the right, is connected directly to the accelerator pedal, but the secondary throttle, left upper, only begins to open when the primary throttle is more than half open. To prevent stalling should the accelerator be snapped open at low engine speeds an automatic valve is located below the secondary throttle. This automatic valve or butterfly is held closed by a counterweight and will only open at a certain designed induction depression. Under certain conditions fluttering of a sensitive flap-valve of this type might occur and the makers have provided a simple oil-filled damper (rather like the S.U. piston damper) to control the movement of the automatic valve.

The standard Solex accelerating pump is used with this carburettor, but a special feature is that operation of the pump can be modified by adjustment of the linkages to work for the primary throttle alone, the secondary throttle alone, or for both throttles simultaneously. Fig. 48 serves to show the working parts of the standard Solex accelerating pump. When the throttle is closed the diaphragm, M, is forced to the right by the internal spring. This lifts the ball-valve, H, off its seat and the pump chamber fills with fuel. Opening of the throttle moves the lower arm of the pump lever, L, to the right which causes the upper arm to force the diaphragm to the left and to inject the contents of the pump chamber through the pump jet, Gp, and the injector, i. The volume of the injection can be varied by a simple adjustment on the operating rod and the speed of injection can be regulated by the size of the pump jet, Gp.

Solex economy

A variation on the standard accelerating pump is the provision of a small diameter push-rod projecting to the left of the pump diaphragm spindle. In the full-throttle position the rod pushes open a spring-loaded valve in the base of the chamber to permit a continuous flow of petrol to pass through the pump jet Gp and out of the injector. At throttle openings of about 80 per cent or less this spring-loaded valve is closed and the speed jet is out of action. This arrangement permits the use of economy settings for the main and air correction jets to cover cruising conditions, since the speed jet will supplement the main supply to give a maximum power mixture at full throttle.

TUNING THE SOLEX CARBURETTOR

Where a Solex carburettor has been fitted as standard equipment to a completely standard engine the only tuning operation that is usually found necessary is the setting of the idling mixture.

The idling mixture

The idling speed of the engine is controlled by the slow-running screw (not shown in the figures), which is mounted on the abutment plate of the throttle lever. When the engine is at working temperature the slow-running screw should be adjusted to give a fast idling speed of about 700 r.p.m. The volume control screw, W, should then be opened gradually (anticlockwise) until the engine begins to hunt, i.e. vary in speed with a regular rhythm. At this point W should be carefully closed (clockwise), no more than half a turn at a time, until hunting stops. The idling speed is next reduced to about 500 r.p.m. Should this cause a resumption of hunting a slight adjustment of W in a clockwise direction will restore regular idling. This is precisely the maker's directions for single carburettor installations. For multi-carburettor installations a more sensitive method is provided by the vacuum gauge technique described on page 26.

Tuning after modification

Even minor changes in the nature of supertuning can sometimes call for changes in the standard carburettor jet settings. A change in the exhaust system can upset carburation and supertuning of a major character, involving such changes as increased compression ratios, larger valves or modified camshafts, can sometimes upset carburation enough to cause misfiring at certain speeds. Another case that obviously calls for special carburettor tuning is when a non-standard carburettor is to be fitted to an engine, or a change from the standard choke size is to be made.

The golden rule for tuning the Solex carburettor is simple: the main jet influences the bottom end of the power curve, the air correction jet the top. This is an oversimplification, since the main jet still has a little influence on the mixture strength at the peak of the power curve and the air correction jet is not entirely inoperative at low speeds. The procedure in general is as follows. First set the idling mixture. If satisfactory idling cannot be achieved by the methods already described and there is no reason to believe that some other malfunction, such as a leaking induction manifold, is upsetting the idle, try fitting a different size of pilot jet. If, for example, hunting cannot be stopped by screwing in the volume control screw a smaller pilot jet is required. If satisfactory idling is still unobtainable, forget the carburettor for the present and look at the rest of the engine. Check for an ignition system defect, incorrect valve timing, even for a rag stuffed in the induction manifold.

When good idling is obtained a quick road-test soon indicates how close we are to satisfactory carburation. If there is a tendency to hesitate, or misfire, at the low end of the r.p.m. range or even a vague feeling of 'woolliness', fit a larger main jet. If the lack of power is apparent at the top end of the range, fit a *smaller* air correction jet, remembering that a richer mixture is given by a reduction in the air correction jet. Changes in main jet are usually made in small steps, never more than five numbers at a time. Changes in air correction can be made in steps of ten or even twenty numbers. When removing the air correction jet care is necessary or it will drop down the choke tube as far as the throttle-plate. Removal and re-fitting the air correction jet is a much simpler operation if the four square-headed bolts holding the top cover in place are removed first.

A particular example of a tuning routine might be an initial try with:

Main 120; Air Correction 190.

After road-testing it is found that bottom end performance is improved with:

Main 125; Air Correction 190

and is about the same with:

Main 130; Air Correction 190.

Further experiments, such as stop-watch checks on the acceleration time 50-80 in top gear, suggest that top end mixture is too rich and a better performance at the top end is given by:

Main 130; Air Correction 240.

Further stop-watch tests over the whole range lead us to adopt the final settings of:

Main 125; Air Correction 220.

Accelerating pump

The procedure is somewhat different when an accelerating pump is fitted, since a lavish squirt of petrol into the choke tube from an oversize pump jet when accelerating can mask the effect of an undersized main jet. The writer usually avoids this danger by testing bottom end performance on a hill. One method would be to approach a certain marker on the hill (a lamp standard for example) at 30 m.p.h., having been at full throttle for several seconds. The speed achieved when passing a second marker (or the time between markers) is used to find the correct main jet. After this step, the correct pump jet can be found by taking acceleration times, such as 0-50 m.p.h. through the gears, increasing or reducing the pump jet Gp in steps of five and increasing or decreasing the pump stroke. Where the accelerating pump also serves an economy function it is the aim to use a relatively small jet setting for economical cruising and to depend upon the pump jet to provide the extra petrol for full-throttle power. The correct combination of main jet and pump jet cannot always be found by a few quick tests and comparative petrol consumption tests over 100-200 miles are sometimes required before a final choice can be made. It should be remembered that the *quantity* of petrol injected by the pump during acceleration is controlled by the pump stroke, which can be varied by removing the split-pin from the pump lever operating rod and refitting the pin in a different hole. The adjusting nut and lock-nut shown in Fig. 48 are not fitted in the majority of Solex pump-type carburettors.

The size of the pump jet Gp governs the *speed* of injection after the diaphragm has moved. When used to augment the full throttle mixture, the size of this jet also fixes the *quantity* of this supplementary fuel supply. For the perfectionist this dual function of the pump jet can occasionally be conflicting, the ideal setting for acceleration not necessarily being the ideal for a speed-jet. Nevertheless a compromise setting is usually very satisfactory for both purposes.

When fitting a pump-type carburettor to a new installation attention should be paid to the type of injection tube fitted. The injector shown in Fig. 48 is called the 'high injector' and is normally used with four-cylinder engines. The 'low injector' as used on six-cylinder engines projects into the throat of the choke-tube.

The starter carburettor

Tuning the starter carburettor is generally confined to choosing the correct setting for Gs, the starter petrol jet. Black exhaust smoke and hunting immediately after starting indicates too large a setting of Gs. Stalling of the engine immediately after starting is cured by the fitting of a larger petrol jet.

One cannot attempt to cover the tuning variables on all types of Solex carburettor in a book of this size. Messrs Solex are always willing to help with advice on any tuning problem and a leaflet on any type of Solex carburettor can be obtained from the Solex Works, 223-231 Marylebone Road, London N.W.1. In America advice, manuals and spares can be obtained from The Arnolt Corporation, Warsaw, Indiana.

WEBER CARBURETTORS

Perhaps it is a platitude to say that the Weber is the Rolls-Royce of carburettors, but the comparison is really very apposite. It was not genius or even originality of thought that put the name Rolls-Royce on the pinnacle of automobile engineering fame—it was sheer high quality and incomparable workmanship. In the same way Edouard Weber and his associates have used well-tried principles of carburation to produce a beautiful mechanism—a carburettor built regardless of expense.

The air-bleed system of correction, as used in Solex and Zenith-Stromberg carburettors, is used on all Webers. The design of the main jets, emulsion tubes and air correction jets differ widely from type to type. There is little sense of standardisation, but the basic system of metering and correction remains common. The idling system is normal, sometimes with one progression hole, sometimes with two. The accelerating pump is a simple piston-type, mechanically operated by the throttle; starting is either by auxiliary carburettor or by strangler. Almost all Webers use a double choke arrangement, the smaller one, to which the fuel is fed, usually being called the auxiliary venturi.

The auxiliary venturi

The use of a second venturi discharging into the throat of the main venturi is an old idea. It is used on all Stromberg carburettors, both automobile and aero, and is in common use on such American

carburettors as the Carter and the Holley. Some American carburettors even use three venturi. There is undoubtedly a swing
towards the use of multiple venturi on smaller carburettors since
the end of the war and it will be of value here to consider the
reason for this.

Professor Lichty (*Internal Combustion Engines*, published by
McGraw-Hill) gives us the reason in a few words. 'For the same
overall loss in pressure, the pressure drop obtained in the boost
venturi is as high as double that obtained in a single venturi.'
(Boost venturi is another name for auxiliary venturi.) Let us expand
this a little for the non-technical. The purpose of the choke or
venturi is to provide a depression (pressure drop) to induce a flow
of fuel into the air stream. The greater the pressure drop, the
greater the flow of fuel and the greater the degree of atomisation.
The price paid for this pressure drop at the throat of the venturi
is a reduction in pressure in the induction manifold and a corresponding drop in volumetric efficiency. Thus the arrangement of
venturi that gives us a certain pressure drop applied to the fuel
feed nozzle for the lowest overall loss of pressure is the most efficient
from the power-producing point of view. Except for the very
smallest of carburettors, the most efficient arrangement has been
proved by laboratory experiments and by test-bed work to be the
double arrangement. Experiments by F. C. Mock in 1942 (S.A.E.
Trans., 50, 105) on single and double venturi showed that the
double arrangement can give nearly twice the metering suction for
the same pressure loss at low and medium air flows and at least
50 per cent more at high air rates. An interesting side-light on
aircraft applications is that the beneficial influence of the second
venturi is reduced at altitude, and at high air-flows and high
altitudes the single venturi is the more efficient. This then is the
evidence in support of the double venturi and one reason for the
efficient operation of the Weber carburettor.

A simple explanation for the higher efficiency of the double
venturi is offered in Fig. 49. It is well known that the velocity of
a stream of air in any conduit is always higher at the centre. The
distribution of velocities across the section varies, depending upon
the type of flow, stream-line or turbulent, but in all cases the
central core of air is travelling at a higher average velocity than
the main stream. Thus with the auxiliary venturi we use the

'boosted' velocity of the central core to give a higher metering suction than we are able to obtain from the single venturi.

The gain from the use of a double venturi can be applied in two ways. It can be used to improve bottom end torque without loss of power at the top end, or it can be used to give the same bottom end

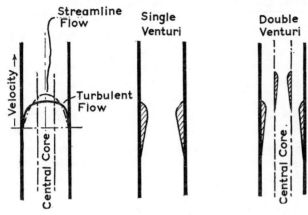

FIG. 49

torque as the single venturi with improved torque and power at high speeds. In the case of the Weber carburettor, especially when fitted to a racing or sports car, it is customary to concentrate on top end performance.

The Weber Company are not given to oversimplification, either in design or in range of models. There are downdraught, horizontal and even updraught models. There are single-choke, twin-choke and four-choke types. There are also twin-choke types with differential opening of the two throttles (as used on the Citroen DS19). A description of the well-known DCO type will serve as an introduction to a typical Weber layout.

Weber Type DCO

A series of carburettors of this general designation have been produced for many years. This series has seen its greatest application in Grand Prix cars. The DCOE series is very similar in layout, but with the addition of a progressive starter device, rather like the Solex Progressive Starter. The DCO series are horizontal twin-choke instruments incorporating air trumpet entry scoops, centrally disposed emulsion tubes and jets, accelerator pump and a high speed automatic device (power jet). Apart from the common float chamber with twin floats arranged to straddle the central jet system (see Fig. 51), all other components are duplicated and the instrument functions as two separate carburettors, with separate metering systems, chokes and throttles. Two such instruments used on a four-cylinder engine thus provide a separate carburettor to each cylinder. With such an arrangement it is possible to take advantage of the induction pulsations by 'tuning' the overall length of the induction system to boost the maximum power of the engine, or for certain circuits to produce a 'hump' in the power curve at medium r.p.m. to improve acceleration. Short, medium and long air trumpets are available for this purpose. Occasionally the pulsations in the induction system are so powerful that the outward-travelling pulse ejects a spray of fuel at the mouth of the trumpet at certain engine speeds. To help reduce fuel loss from this cause Weber provide extension tubes (20 in Fig. 50) which can be fitted to the auxiliary venturi.

An original feature of this carburettor is the use of twin floats to nullify the effects of fuel surge during cornering. It will be seen from Fig. 51 that the main jets and emulsion tubes (7) are centrally placed between the two floats. Inclination of the fuel surface under cornering, accelerating or braking forces will have very little effect on the head of fuel applied to the main jets.

Fig. 50 is a sectioned view of the type DCO carburettor. It is not an orthodox cross-section, several items being displaced from their true position for the sake of clarity. The air entry trumpets are omitted.

Idling

Fuel for the idling jet, 10, passes from the float chamber through the side holes in the communication bushing, 19. Fuel is metered through the calibrated hole in the base of the jet body and passes out through the side holes to mix with air from the idling air screw, 11. The small passage, 18, carries the emulsified mixture to the idling mixture hole, 30, and the progression hole, 29. The volume control screw is shown at 32.

To the right of Fig. 50 is shown an alternative arrangement found on certain DCO type carburettors. In this a longer emulsion tube, 35, is used and the bushing, 19, is omitted. The feed to the idling jet is thus through the main jet and then through the side holes in the emulsion tube. In certain conditions this second arrangement can sometimes be inadequate. For example: after a

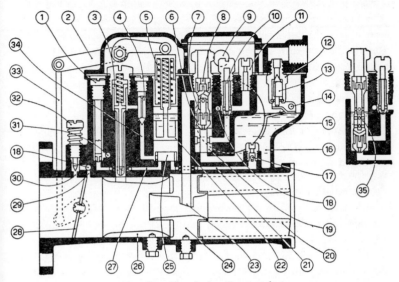

FIG. 50.   Descriptive Cross-section

1. Pump exhaust screw; 2. External pump control lever; 3. Pump jet; 4. Needle delivery valve; 5. Piston return spring; 6. Main jet; 7. Emulsioning tube; 8. Emulsioning tube air adjusting screw; 9. Air scoop connection; 10. Idling jet; 11. Idling air screw; 12. Needle valve seat; 13. Needle valve; 14. Float fulcrum screw; 15. Float; 16. Float bowl; 17. Pump intake valve; 18. Idling mixture tube; 19. Idling communication bushings; 20. Auxiliary Venturi extension pipe; 21. Pump intake tube; 22. Pump piston; 23. Discharge tube; 24. Auxiliary Venturi; 25. Spacer for piston stroke reduction; 26. Choke; 27. Pump exhaust tube; 28. Throttle; 29. Progression hole; 30. Idling mixture hole to the intake pipe; 31. Pump exhaust tube; 32. Idling mixture adjusting screw; 33. Pump control shaft; 34. Pump delivery tube; 35. Emulsioning tube.

period on full throttle the well (the annular space around the emulsion tube, 7) is drained of fuel. Sudden closing of the throttle puts the engine back on the idling system, but the well must fill with fuel, in the second arrangement, before fuel can pass to the idling jet. Persistent cutting out on dropping back to idling speed can be caused, of course, by an incorrectly tuned idling mixture, but the limitations of the second feed arrangement should not be forgotten.

Main metering system

The main metering system is all centred in one component, the emulsion tube, 7 (or 35). The air correction jet, 8, and the main jet, 6, are fitted to the top and bottom of the emulsion tube and are locked in position by the hexagon-headed tube holder (7 in Fig. 51). The emulsified mixture passes into the vertical discharge tube, 23, and enters the air stream at the point of maximum depression in the auxiliary venturi. The degree of emulsification will be greater when the long tube, 35, is fitted. It will be seen that both main and idling circuits draw their air supply from the jet access chamber, which in turn is vented to the top of the float chamber. The jet access chamber is vented to atmosphere through the port, 9. When the carburettor intakes draw their air supply from an air-box pressurised by forward-ram, it is necessary to take a pipe from this port to a point on the air-box.

Accelerating pump

Accelerating pumps on Webers are always of the piston-type and on this particular model a novel feature is the provision of a 'prolongation spring' to regulate the rate at which the piston descends. In Fig. 50 it will be seen that a spring-loaded plunger is interposed between the pump control lever, 2, and the pump piston, 22. The outer spring serves to keep the plunger in contact at all times with the control lever. The inner spring is the prolongation spring which is compressed to a greater or less extent depending upon the speed of opening the throttle. Thus for a slow opening rate no compression takes place and the pump piston descends at the same speed as the plunger. A quick stamp on the accelerator compresses the spring and the rate of injection is controlled by the recovery rate of the inner spring, not by the speed at which the throttle is opened. Closing of the throttle causes the piston to rise under the action of the outer spring, filling the pump chamber with fuel drawn from the float chamber through the intake valve, 17. Fuel is delivered to the two chokes through calibrated pump jets situated towards the rear of each main choke. It should be noted there is only one accelerating pump serving two pump jets.

The pump stroke can be varied by inserting spacers of differing height at the base of the pump chamber. The amount of fuel injected can also be varied by changing the exhaust screw to one with a different size of drain hole. This, screw, 1, has in its tip a calibrated drain hole to lead a certain amount of the pump delivery back to the float chamber. On certain installations this screw is undrilled. The rate of injection is controlled by the size of the pump jets.

High-speed device

The needle valve, 4, besides serving as a non-return valve in the pump delivery circuit, is designed to act as a power jet. When the depression created in each main choke at the tip of the pump jet has reached a certain value, depending upon the weight of the needle valve, the valve is sucked off its seat and suction is applied through the pump chamber and the intake valve, 17 to draw fuel from the float chamber and inject it at the two pump jets. This only occurs at high engine speeds and full-throttle and serves to enrich the mixture and prevent overheating of the engine under arduous race conditions. The size of the pump jets control the amount of this supplementary fuel and the weight of the needle valve determines the point of opening. In the type DCOE a steel ball held down by a tubular weight is used instead of a needle valve.

On other types of Weber carburettor a different device, called a full-power device is provided. This is throttle operated and between two-thirds open and fully open a valve is held open to admit a metered supplementary feed. The high-speed device only enriches the mixture at high engine speeds, but this device operates whenever the throttles are wide open, even at low engine speeds.

Tuning the Weber carburettor

The setting of the idling mixture on the Type DCO carburettor is carried out in the usual manner as described for previous fixed choke carburettors. There are two mixture adjusting screws (14 in Fig. 51), one to each idling circuit, but the extent of opening of the two throttles is adjusted by one throttle-stop screw (9 in Fig. 51), since the two throttle plates are mounted on a common spindle.

Small changes in the sizes of the main and air correction jets to suit the particular barometric pressure, humidity and temperature are all that are necessary in general to tune Webers to a given circuit. Access to the jets is given by removal of the small cover in the centre of the float chamber cover. When it comes to the choice of chokes to suit different circuits even the professionals find it difficult to make up their minds. When one has two chokes to play

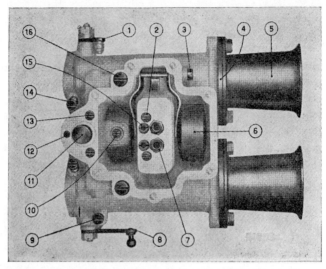

FIG. 51. At request, the carburettors of the '35-38-39-40 DCO' type can be supplied completed with the auxiliary Venturi extension pipes (4) and additional air horns (5) having the shape of a trumpet; moreover a universal arrangement of the main jets, emulsioning tubes and float allows the carburettor to be fitted either on the right or on the left engine side assuring a steady feed under all conditions. In order to indicate the position of the internal pieces of the carburettors of the '35-38-39-40 DCO' type, a plan view without the cover is shown in Figure 51.

1. Pump control main lever; 2. Idling jets; 3. Float fulcrum screw; 4. Auxiliary Venturi extension pipes; 5. Additional air horns; 6. Float; 7. Emulsioning tubes complete with main jets and air adjusting screws; 8. Throttle control lever; 9. Idling speed adjusting screw; 10. Pump intake valve; 11. Accelerating pump; 12. Pump exhaust screw; 13. Pump delivery valves; 14. Idling mixture adjusting screws; 15. Idling air screws; 16. Pump jets.

with one can ring so many changes on sizes. It is obvious that an improvement in low-speed torque can be expected from a reduction in size of both main and auxiliary, but it requires experience of *both carburettor and engine* before one can predict with any accuracy the effect of:

> (*a*) a small reduction in main choke, accompanied by an increase in auxiliary

or (*b*) a small increase in main choke, accompanied by a decrease in auxiliary.

It should be noted that the outside diameters of Weber auxiliary venturi are all the same size. It is the size of the throat that is changed.

Obviously, every change in choke or venturi size necessitates changes in main and air correction jets to maintain the correct mixtures throughout the range. One must know, therefore, before practice starts, which jets go with which chokes. All this preliminary work is best carried out on the dynamometer, but the poor amateur can only try to make up this deficiency by the expenditure of much time and infinite patience.

Should it be necessary to tune the accelerating pump to a new installation, it should be remembered that the rate of injection can be increased by fitting two larger pump jets, but the total amount of fuel injected can be increased in two ways, either by removing the spacer (25 in Fig. 50), or by fitting a pump exhaust screw with a smaller bleed hole. The choice of pump jet on a racing installation, is largely governed by the extent of enrichment required at high engine speeds and full-throttle. If the size of pump jet to satisfy this requirement exceeds the minimum requirement for good acceleration the drain hole should be increased to reduce the quantity of the injection when accelerating.

Weber carburettors are rather costly and there is a tendency for the amateur race competitor to try to obtain a long life from his Webers, sometimes transferring them to different engines, or even trading them to a fellow competitor for the price of a new pram. Wear inevitably takes place in the throttle spindle bosses, at the throttle edges and in the barrels. Patient work with fine emery cloth can remove the ridges in the barrels made by the throttle edges and the fitting of new throttle plates will in general complete this particular reconditioning operation. Sometimes, however, the increase in diameter is sufficient to alter the position of the throttle edge relative to the progression hole (or holes). This can delay the opening of the progression hole, resulting in a flat-spot. A simple expedient to overcome this defect is the filing of a small chamfer on the downstream edge of the throttle-plate. This should be carried out in stages, only removing a small amount of metal at a time, until the flat-spot disappears.

When fitting new throttle-plates to Webers care should be taken to see that the replacement plates are of the correct angle. Plates are marked as 85° or 87° and should not be interchanged.

Practical and technical assistance and spare parts for Weber carburettors can be obtained in Great Britain from Fiat (England) Ltd, or direct from Edouardo Weber, Via Timavo 33, Bologna, Italy. In America: *West Coast*, Carveth Enterprises, 770 El Camino, San Carlos, Calif.; *East Coast*, GEON, Great Neck, New York.

CHAPTER SEVEN

Testing and Tuning

BEFORE TUNING one should always test. This should be an inflexible
rule. Sometimes one is asked to tune an engine which is obviously
running badly, misfiring, or lacking in power. If the fault is not
immediately apparent there is only one possible course: carry out
a series of tests to reveal the cause or causes of the bad running.
But what of the engine that appears to be running well? Herein
lies the temptation. Why not dispense with testing and go straight
into the business of tuning? Because testing is like a medical
check-up—it sometimes discloses weaknesses of which we were not
even aware. Here are two examples. The compression tester can
reveal incipient valve failure, even when the seat leakage is still
too small to have any noticeable effect on the power developed by
the engine. The condenser tester can reveal the breakdown of the
insulation inside a condenser long before it has any apparent effect
on engine performance. Testing then is a necessary preliminary to
tuning, if we are to have any confidence in our system of maintenance.

When an engine is being tested to disclose the source of a fault
or a misbehaviour of the engine, such as difficult starting or misfiring at speed, there is a great temptation to abandon the remainder
of the schedule of tests when a defect has been discovered that is
consistent with the symptoms. Experience shows however, that
only about 70 per cent of engine troubles come singly. Two, three
or even four defects may all be contributing to the symptoms.
Only by insisting on the completion of the whole series of tests,
checking compressions, fuel system and ignition system, can we
hope to know the existing state of tune of the engine. Later, if our
efforts at tuning the engine still leave us with an engine that is not
up to standard for the particular model, there are more tests we
can carry out to help us to diagnose the trouble. These tests are
discussed under the heading of Trouble-Shooting in Chapter
Eight. For example, on an engine that cannot be tuned to give a

satisfactory vacuum reading, one would check the valve timing to see if this had been incorrectly set. Such a test is not included in this present list of tests, since valve timings do not usually change without human intervention. The pre-tuning tests described here are tests to disclose defects caused by wear and tear, such as leaking valves, pitted contact breaker points, or a worn petrol pump.

The order of precedence for the tests is not important and the order given below can be varied to suit individual cases.

TESTING

Test No. 1. Road-test

Whenever possible one should carry out a short road-test of a car before testing and tuning. Nothing is more convincing to the layman than a practical demonstration and the tester must be certain that his efforts have improved the car's performance before trying to demonstrate the improvement to the customer.

Ideally one should cover a whole range of acceleration and speed tests in the manner of a *Motor* or *Autocar* road-test. More realistically, since time is money, an experienced tester usually develops his own test in keeping with his local environment. In the centre of a large town or city it is difficult enough to find a suitable quiet stretch of road to time accelerations, even between such a low range as 10 to 30 m.p.h.

For safety's sake the author usually takes an assistant with him to operate the stop-watch and usually concentrates on two acceleration tests, each test being repeated until he is satisfied it is representative. Acceleration tests of this nature never start from rest. No customer is pleased to learn that one of his axle-shafts has been snapped by standing-start tests to prove how well his engine has been tuned.

For a car such as the Triumph TR3 or TR4 two suitable tests would be:

(a) acceleration time 20 to 40 m.p.h. in top gear.
(b) acceleration time 20 to 40 m.p.h. in second gear.

By covering the same road speed in these two gears the torque curve is tested in two sections. Test (a) covers the torque range from approximately 1000 to 2000 r.p.m. (in the case of the TR3); the second test covers the torque range roughly from 2000 to 4000 r.p.m. If our tuning has improved one part of the torque curve at the expense of the other these two tests will show it up.

For a quick check on performance the writer sometimes uses an acceleration test, through the gears, up a fairly steep hill. The test starts by crawling in bottom gear up to the starting point, then doing a speed hill climb to the final marker. Such a test on a short section of a long straight hill will yield as much information as an acceleration test from 5 to 100 m.p.h. on the level. If a suitable gradient is available one need not exceed 50 m.p.h. in a hill-climb test. In this way one is less liable to incur the displeasure of police patrolmen.

Test No. 2. Compression tests

It is useful, when carrying out these tests, to know what compression pressures one should expect to find on the particular model when in good order. These figures are sometimes given in service manuals. The instructions for carrying out the compression tests will also be found in the service manual and particular attention should be paid to one point. Some makers recommend that the throttle should be *closed* while the compression pressures are being measured; others recommend that it should be *open*. With the throttle closed slightly lower readings are given. If the maker's manual does not supply typical compression figures, much can still be learned from a comparison of the different readings from cylinder to cylinder.

It is essential that the engine be at working temperature before these tests are performed. If the road-tests have just been carried out the engine will be already at working temperature. Since plug tests can conveniently follow the compression tests the ignition should be cut immediately upon the return from road-testing. This will leave the evidence of the full-load temperatures of the plugs in their colour and general condition. If the engine is allowed to idle as we enter the garage this useful evidence will be covered by a film of soot.

The procedure for compression testing is as follows: The sparking plugs are removed and put on one side, in the correct cylinder order, for examination later.

Dry compressions

The compression gauge is fitted to each plug hole in turn and the engine turned over on the starter. In the case of the Crypton instrument it is necessary to press the rubber adapter fairly tightly into the plug hole to achieve a seal. This is a two-handed job and an assistant is required to press the starter button.

It is quite easy to count the number of engine revolutions performed under the action of the starter, since the engine is turning against one compression only, the other plug holes being open. This gives a distinctive 'beat' to the noise of the engine turning. The assistant operating the starter button should carefully count the number of revolutions and take his finger off the starter at the eighth revolution. The compression gauge reading is taken and recorded on the test report. The non-return valve in the end of the instrument is opened and the needle brought back to zero. This procedure is repeated until dry compression readings are obtained from all cylinders.

Wet compressions

Approximately one tablespoonful of oil (S.A.E. Grade 10 or 20) is poured in the plug hole of No. 1 cylinder. This can be conveniently performed by means of a pump type oil can, the number of pumps required being measured first. The compression test is repeated, turning over the engine eight times on the starter as before and the 'wet' reading recorded on the test report on the line below the 'dry' readings. A similar quantity of oil is now added to No. 2 cylinder and the procedure repeated until a full set of wet readings are obtained.

The purpose of the oil is to provide a temporary seal to the leakage of air past the top ring of the piston. Thus if all the valves are in good condition and all the piston rings are in good order on three cylinders, but are broken on No. 2 piston, we would expect to find a set of readings of the following pattern:

	1	2	3	4
Dry Compressions, lb. per sq. in.	130	102	128	129
Wet Compressions, lb. per sq. in.	140	141	138	140

Alternatively, if all the piston rings and bores are in excellent condition, but No. 2 exhaust valve is beginning to burn, we would expect readings as follows:

	1	2	3	4
Dry Compressions, lb. per sq. in.	130	90	128	129
Wet Compressions, lb. per sq. in.	140	101	137	141

A useful rule then to evaluate valve and cylinder condition can be stated:

(a) *Cylinder condition*

Subtract dry readings from wet readings—the closer the readings the better the cylinder and piston ring condition. Differences of less than 10 lb. per sq. in. are excellent. Differences exceeding 30 lb. per sq. in. are poor.

(b) *Valve condition*

Take the highest wet reading to be a perfect seal and subtract the other wet readings from this one to assess the effectiveness of the valve seal in each cylinder. A difference of no more than 5 lb. per sq. in. indicates a good valve. A difference of 20 calls for a repeat check on the wet reading from this cylinder. A difference of 40 lb. per sq. in. is a sure sign of a bad valve.

Interpretation of readings

There are several pitfalls to avoid when interpreting compression tests. The first is the meaning of 'cylinder condition'. The ability of a piston and its rings to seal against gas leakage in a particular cylinder bears no relationship to the ability of the same piston and rings to control the passage of oil into the combustion chamber.

One will occasionally find an engine that gives dry readings within 15 lb. per sq. in. of the wet readings, yet this engine is known to burn excessive quantities of oil. One can also sometimes find an engine where the reverse takes place, i.e. poor gas sealing with low oil consumption. The compression tests give no indication of the oil usage of an engine.

A second warning is necessary when a side-valve engine is to be tested. The wet test is not at all reliable in the case of this design

SCE K

of engine. Since the plug hole is not directly over the cylinder it is difficult to spread oil over the piston crown by squirting it through the plug hole. On some cylinders the operation might be successful, on others it might not and misleading conclusions could be drawn from the figures obtained. With side-valve engines only the dry readings should be taken. Any variation greater than 30 lb. per sq. in. points to a serious leakage past a valve seat.

A final warning is necessary on the subject of cylinder head gaskets. A leaking head gasket which permits the passage of gas outwards, can be heard when the engine is running by the noise of the escaping gas. There is little danger in this case of the tester mistaking a low compression reading from this cause for a badly seating valve. However, when low compression readings are given by two adjacent cylinders one cannot differentiate between the two kinds of failure. Two adjacent exhaust valves could be failing or the gasket between the two cylinders could be leaking or burned.

Test No. 3. Plug tests

Much can be learned from the appearance of the plugs after running on full load, In the absence of an exhaust gas analyser, this knowledge can be a valuable guide to the mixture strength provided by the carburettor. Before judging the mixture strength from the plugs, however, one must be satisfied that the plugs are of the correct heat value for the particular engine. If the engine is not modified in any way from standard and the plugs fitted are those recommended by the engine manufacturer (or a different make of plug of equivalent heat value and reach) one can use the colour and state of the plug points and the nose of the insulator as a guide to mixture strength under load. If the compression ratio has been raised or the engine is supertuned in any way, the evidence of the plugs requires a different interpretation.

For the standard engine with correct plugs the following information is sound:

(a) an insulator nose that is dry, clean and off-white in colour indicates an over-weak mixture.

(b) an insulator nose that is browny-grey or fawn in colour and is fairly clean and dry indicates that mixture strength is not far from the correct one.

(c) a black, sooty condition indicates an over-rich mixture.

(d) an oily condition indicates excessive oil consumption. This oil may be passing the piston rings or being drawn down the inlet valve guides on the induction stroke. With an oily plug no indication of mixture strength is given.

With a non-standard engine white plug points can also indicate that the plugs are too 'soft' for their duty. Similarly, sooty plugs can indicate that the plugs are too 'hard', i.e. are of too high a heat value for the particular engine.

The plugs should also be examined at this stage for cracked insulators or for signs of spalling of the material around the centre electrode. Finally the gaps are checked and re-set to the recommended gap-setting. Any great discrepancy between the existing gaps and the recommended values should be noted on the test report, since an abnormally small gap can cause difficult starting and an abnormally large one can be the cause of a high speed misfire.

Test No. 4. Initial vacuum reading

The vacuum gauge should be connected to the induction manifold in the manner described in Chapter Two.

The existing vacuum reading of the untuned engine can be a guide to the state of tune. For a typical modern sports car engine a reading between 17 and 21 inches is normal at a fast idling speed of 700-800 r.p.m. For the reasons stated earlier in Chapter Two the idling vacuum readings do not serve as a comparative guide to the efficiencies of different designs of engine. Many low speed touring type engines give higher vacuum readings than high speed, high compression competition engines. For a given engine type, however, the initial vacuum reading is a useful measure of the state of tune. A Triumph TR3 in a good state of tune will give a vacuum reading of 20-21 inches. If the pre-tuning vacuum reading is only 17 inches the road performance of the car will be noticeably below par.

The reading should be fairly steady. The degree of permissible fluctuation cannot be laid down since it is influenced by the degree of damping in the vacuum instrument and the length and bore of the connecting piping. A pronounced 'kick' downwards at intervals can be an indication of a burnt or sticking valve or a leaking cylinder head gasket. The compression readings will have already shown if any cylinder appears to have badly-seating valves or a gasket leak.

Test No. 5. Battery condition

The specific gravity of each cell should be checked in turn. A battery in a good state of charge should give readings between 1·300 and 1·220 at 60°F (1·310 to 1·230 at 85°F and 1·290 to 1·210 at 35°F).

If the average gravity for all cells falls below the specified minimum the battery is in a discharged condition and should be recharged. Any variation of more than 25 gravities (0·025) between cells indicates a defect in one or more cells. A closer examination by a battery service man is advisable in such a case. The use of the hydrometer is not always convenient, especially when accessibility is poor and removal difficult. For such cases the Crypton 'Hydro-lek' meter is much more convenient. This instrument measures the open-circuit voltage of each cell. The variation in this voltage between a fully charged and a fully discharged cell is very small, but the 'Hydro-lek' meter can measure this voltage with sufficient accuracy to make it a reliable guide to the state of charge in a battery.

Test No. 6. Voltage at coil and starter

Having established that the battery is in good condition and capable of maintaining its correct voltage for a reasonable period of time, the next stage is to see that this voltage is available at the coil and at the starter.

To test the voltage at the coil the voltmeter is connected from the S.W. terminal on the coil to earth. The voltage measured in this way should be almost the same as full battery voltage. A drop of more than 0·5 volts indicates an abnormally high resistance in the leads or terminals. In such cases a quick check over the wiring usually discloses a loose or dirty connection. If the wiring and all connections appear to be sound the voltage drop across the ignition switch should be measured to see if the high resistance is seated here.

The voltmeter should now be connected across the starter. With a positive earth system this means that the negative lead from the voltmeter should be connected to the main starter terminal and the positive lead clipped to earth. The starter button should be held closed for about 10 seconds while the voltage reading is observed. For a 12 volt system the reading should not fall below 10 volts. In cases where a greater voltage drop than 2 is observed all connections should be checked, also the conditions of the earth strap. If these are sound the starter should be removed for checking by a competent auto-electrician.

Test No. 7. Coil and condenser tests

Using the Crypton coil tester or a similar instrument, the H.T. windings are checked for continuity. A chart is provided with the Crypton instrument giving the normal stall current to be expected from different types of Lucas coil. A reading much higher than the normal for the particular type of coil means a lower than normal secondary winding resistance, which in turn suggests that a section of the windings are shorted out. If the reading is too low, the resistance must be too high, possibly the result of internal corrosion or loose internal connection. In either case, a resistance too low, or resistance too high, it is advisable to fit a new coil.

If no reading is given, after checking for poor connections at the terminals, it must be taken that the primary winding is open-circuited. The temperature of the windings affects the readings obtained. The readings on the chart provided with the Crypton coil-tester show an acceptance band based on a cold coil. When tested with the windings at working temperature, the coil output should not fall into the 'bad' zone of the scale.

Complete failure of the condenser will make the spark so feeble that the engine will refuse to run. Partial failure will produce a misfire.

Three tests are necessary to establish that a condenser is serviceable. These three tests are for capacity, series resistance and leakage. When using the Condenser Tester the lead from the coil (the C.B. terminal) is first disconnected and the contact breaker points are held apart by a match-stick. One lead from the Tester is connected to the C.B. terminal on the distributor. The other is clipped to earth on the distributor body. With the typical British four or six cylinder engine the condenser capacity falls within the limits 0·18 to 0·25 microfarads. When checking the series resistance of the condenser it is advisable to measure it in two ways. A high resistance reading when the condenser is in position could be caused by a loose connection. By disconnecting the condenser and measuring its resistance separately this possible source of additional resistance can be eliminated. Similarly, when testing for leakage, a condenser should first be tested *in situ*. If a certain amount of leakage is indicated, the condenser should not be condemned until a second test has been made with the condenser removed from the distributor and the test leads connected to the condenser terminal and to the outer case of the condenser. If leakage is still shown, the condenser insulation is at fault. If however, the second test shows no leakage, and the first has shown some, it is advisable to look to the insulation of the condenser in the distributor. The leakage path can often be traced to dirt or a film of moisture on the baseplate.

Test No. 8. Contact breaker tests

With the contacts in the closed position, the low reading voltmeter (0-1 volts) should be connected across the points. If the points are dirty or pitted the voltage drop will exceed 0·1 volts. Before refacing the contacts, however, the existing dwell angle should be measured. If a tach-dwell meter is not available, the gap can be measured by feeler gauge. This cannot be measured with any accuracy with a 'hump' on one point and a 'pit' on the other, but a rough measurement of the existing gap should always be made if one is to accumulate evidence to account for any misbehaviours of the engine prior to tuning.

Test No. 9. Fuel pump condition

When a mechanical pump is fitted it is advisable to test its capacity immediately after plug testing and before the plugs have been replaced in the cylinder head. This reduces the drain on the battery during the 30 to 40 seconds of starter operation required for this test.

British fuel pumps are confined to two types, the S.U. electric and the A.C. mechanical.

The S.U. pump

The low-pressure, or type L, pump (the type usually fitted under the bonnet) should give a delivery pressure of about $\frac{3}{4}$ lb. per sq. in. when in good condition. The pumping rate of an L type in good condition with an open-ended delivery is approximately 7 gallons per hour, or about one pint per minute. Here we mean the imperial pint. The American half-quart is equal to five-sixths of the imperial pint. The pump delivery rate can be checked against the stop-watch by disconnecting the delivery pipe to the carburettors and inserting the open end of the pipe into a one-quarter or one-half pint measure. An assistant is required to switch on the ignition.

The tester starts his watch immediately the fuel begins to flow, and stops it when the measure is full, at the same time telling the assistant to switch off. An output of a quarter-pint in 30 seconds is the lower limit of acceptability. To test for air leaks in the suction line between the pump and the tank the delivery pipe should be immersed below the level of the petrol in the measure. If, when the pump is working, air bubbles are seen to be rising from the delivery pipe, the pipe line and its connections all the way back to the tank should be checked for the source of the air leak.

If the output of the pump is below the acceptable figure of a pint in two minutes and no sign of an air leak in the suction line is apparent, the pump should not be blamed until the line has been checked for abnormal resistance to flow. A quick check for a partial blockage in the line can be made by blowing down it. Very little effort should be required to produce a bubbling noise in the tank. If the tester finds himself going red in the face with the effort he should suspect a flattened pipe or a blockage. If the poor pump output is finally traced to the pump, the pump should be overhauled, or if more convenient, replaced by a works exchange unit.

The high-pressure or type HP pump (the type usually fitted near the tank) normally provides a delivery pressure of $1\frac{1}{4}$ to $1\frac{1}{2}$ lb. per sq. in. An acceptable lower limit is 1 lb. per sq. in. The normal output is 10 gallons per hour or 1 pint every 45 seconds. An acceptable lower limit for delivery is 1 pint every $1\frac{1}{2}$ minutes. The high pressure pump is provided with a light compression spring to hold the inlet valve on its seat. This acts as a safeguard against the siphoning of fuel to the carburettor when the car is standing. The

LCS type is almost identical to the HP, but the normal output is slightly greater, 12½ gallons per hour.

Note

Never test the current flow to an S.U. pump by disconnecting a lead and sparking across to the terminal. If fuel vapour is hanging around the pump an explosion could occur.

The A.C. mechanical pump

To measure the delivery pressure of this type of pump the engine must be running. The engine will run long enough at a slow idling speed on the petrol in the float bowl while a check is made on the delivery pressure. The pressure gauge should be connected to the pump outlet connection exactly as for the S.U. pump. The normal pressure is 1½-3 lb. per sq. in. To check output, the engine should be turned over on the starter, preferably with the plugs removed, while the pump delivery is measured into a quarter-pint container. A pump delivering less than a quarter-pint in 40 seconds should be overhauled if no indication of air leakage or blockage can be found in the pump suction line.

Test No. 10. Mixture strength

If an exhaust gas analyser is available the mixture strength should be observed (*a*) when idling and (*b*) at a steady speed of 4000 r.p.m. If accelerator pumps are fitted to the carburettors (or dampers in the case of an engine with S.U. carburettors) the throttles should be opened quickly and the analyser needle watched to confirm that a momentary enrichment of one to two ratios does in fact occur several seconds after the throttles have been opened. The time lag before the 'rich dip' is registered on the instrument will, of course, depend upon the length of tubing between the exhaust pipe and the instrument.

The idling mixture strength should be between 12·0 to 1 and 13 to 1. The mixture strength at speed will be rather weaker, around 13 or 14 to 1. In some instances the mixture strength at speed is no weaker than at idling. Neither of these readings, idling or at speed, will show the working of the economiser diaphragm which is fitted to certain carburettors. This device is only in operation at medium engine speeds and with a high induction vacuum. A road-test with the analyser carried inside the car on an assistant's knee will be necessary if the tester wishes to obtain direct evidence of this. In any case the Cambridge, Crypton and Sun Gas Analysers cannot measure mixture strengths weaker than Chemically Correct (about 14·5 to 1 with a typical modern petrol).

Without instrumentation one can still get a good indication of the mixtures provided by the carburettors:

(a) by examining the sparking plugs, as already discussed earlier in this chapter,

(b) by listening to the exhaust note, and

(c) by noting the effect of a temporary weakening or enrichment of the normal mixtures.

(a) The colour and state of the plug insulator nose is a good guide to mixture strength only when the engine is perfectly standard in its specification. If, for example, a previous owner has increased the compression ratio without changing the plug type, he may have fitted larger main jets to the carburettor to overcome the subsequent pre-ignition when under load. Thus the mixture might now be too rich for maximum power, yet the colour of the insulator will indicate a correct mixture strength.

If there is any doubt on this score, the compression pressures already measured should be compared with the values given in the maker's handbook for a standard engine in good condition. An increase above standard of 10 lb. per sq. in. can be ignored, but an increase of 20 lb. per sq. in. or more means that the compression ratio has been raised enough to make a change in plug hardness advisable. No action at this stage is advised but the question of plug heat value will require investigation before the final tuning is completed.

If the engine can be run up to full speed from idling speed, both gradually and quickly, without hesitation or misfiring, no further action is necessary at this stage. On the other hand, if a misfire or hesitation is noticed at any particular speed, the mixture should be enriched, either by pulling out the choke control slightly or, in the case of the S.U. carburettor, by screwing down the mixture adjusting nut four or five flats. The effect of this enrichment on the misfire should be noted. An improvement indicates that the misfire was caused by a mixture weakness, a deterioration shows that the mixture was too rich originally. With this information noted on the report sheet, the carburettor setting should be returned to its original position.

Test No. 11. Valve clearances

It can be assumed in a book of this type that the reader is familiar with the normal method of checking and re-setting valve clearances on the typical push-rod o.h.v. engine. It is convenient to combine testing and tuning in this instance, the clearances, first being measured and then, if necessary, adjusted to the correct maker's figures. The existing clearances on the untuned engine can sometimes throw useful light on deficiencies in performance. For example, insufficient valve clearance can sometimes increase valve overlap at T.D.C. to such an extent that fuel economy is sacrificed. On some engines an increase in valve clearance has been found to give more power at the cost of increased mechanical noise. This will be discussed more fully later. Too fine valve clearances are dangerous. The majority of the heat given up to the exhaust valve heads is conducted away via the valve seats and any failure to make an adequate contact with the seat in the cylinder head, will inevitably result in a burned exhaust valve. An inlet valve will withstand this treatment for a much longer period, but power is lost by leakage during the compression and firing strokes and, when the leakage path becomes large enough, flames escape into the induction pipe, resulting in what is commonly called spitting or back-firing in the carburettor.

Measuring the valve clearances on twin-overhead camshaft engines such as the Aston Martin and the Jaguar is fairly simple. Adjusting the clearances, however, is a laborious business.

TUNING

Opinions differ as to the correct order in which one should carry out the different operations during a tune-up. This is a matter of some importance. If an engine is badly out of tune and is particularly sensitive to both carburettor tune and ignition tune one might easily fail by a significant amount to strike the optimum settings of carburation and ignition at the first attempt, since the two are mutually interdependent. In view of this interdependence it is better to perform the more straightforward operations first, such as adjusting the valve clearances and setting the plug gaps and to leave the more sensitive adjustments till last. If the ignition timing is set before the carburettors are tuned, it is no great chore to re-check the ignition timing, if the mixture has required a drastic change.

The writer has found the following procedure to work out well in the majority of cases:

1. Adjust valve clearances.
2. Clean and re-gap and pressure-test sparking plugs.
3. Adjust contacts.
4. Set ignition timing.
5. Tune carburettors.
6. Road-test.

These six tests will now be considered in detail.

Valve clearances

On the majority of cars this is a simple mechanical operation. Any doubts, however, should be cleared up by consulting the maker's handbook. Some specify that settings should be made cold; others with a warm engine. This should not be ignored. Some modern camshafts have extensive ramps or quieting curves at the beginning and end of the cam profile. These can extend so far into the base circle of the cam that the angular movement over which clearance can be checked with accuracy is very limited. In view of this it is a safe plan to find the point of maximum lift on each cam, then rotate the engine through exactly one revolution (within plus or minus 10 degrees, say) to arrive at the correct setting position on the back of the base circle.

In general the maker's recommended valve clearances are the best. One might expect that a reduction in inlet valve clearance might give an increase in power at the top end of the curve, since this will, in theory at least, give a slightly earlier valve opening, a later valve closing and a slightly greater maximum lift. Oddly enough this is seldom true. It is more usual to find that an increase in valve clearance improves power at the top end. For regular road use this is not to be recommended since it can cause abnormal cam wear, but for use on race day it is well worth a try. But first do a little experimenting to see if it does give an improvement.

Sparking plugs

Our reader should not need the usual warning, or so we imagine! If he has grown careless through familiarity, here it is: when setting the plug gaps, *Don't press against the centre electrode.* Always check on the maker's recommended plug type, if this is not already known. A plug of too low heat value for the particular engine could behave satisfactorily when tested under no-load garage testing conditions. Under load such a plug would soon overheat, leading to pre-ignition, misfiring or rough-running. If the engine is non-standard, with a higher compression, or any special equipment giving an appreciable increase in power it is obvious that the standard plugs are no longer suitable. The selection of the correct type is largely a question of trial. This involves road-testing under full load, followed by examination of the plug insulator noses as described in Test 3 of the Test Procedure. This kind of plug testing and selection does not come within the scope of normal tuning, but the professional tuner, working on a strange car, occasionally meets the problem. Nothing can be taken for granted on a strange car, especially a sports car.

Distributor

The distributor cap should be examined first. The carbon brush in the centre should be examined and the free movement checked. The tightness of the plug leads in the cap terminals is another check. Any signs of neglected maintenance, i.e. excessive dirt in this area, will call for a good cleaning and a more detailed examination. The terminals should be unscrewed and examined for signs of corrosion or cracks in the plastic body. Corrosion or dirt between the stranded cables and the contacts at the base of each terminal socket must be removed. Any signs of persistent leakage of water inside these sockets calls for the application of a liquid jointing compound around the gaps between the rubber covering and the plastic screws.

The contacts should always be removed for examination. If they are badly pitted, new ones should be fitted. A flat fine carborandum stick can be used on old points to recondition them. No trace of hump or pit should remain. In either case, new or old, great care is still necessary to avoid misalignment of the kind shown at (*a*) in Fig. 52 when they have been re-fitted. The second type of misalignment, shown at (*b*) in Fig. 52, is the result of careless fitting after re-facing. A useful and moderately priced handoperated grinding tool, which is specially designed to prevent the

(a) (b)

Fig. 52. Contact misalignment

first type of misalignment, is the Truepoint grinder, which can be obtained from any British garage accessories factor.

After grinding and re-fitting, the voltage drop across the points should again be checked. If this still exceeds 0·1 volts and cannot be improved by careful repositioning of the points, further grinding should be carried out.

While the contacts are removed for grinding it is convenient to remove the baseplate to examine the conditions of the centrifugal advance mechanism. If the pivots of the weights appear worn or the springs appear to have stretched it is advisable to fit new parts. If we are to extract full power from the engine the correct maker's advance curve must be reproduced. This will not happen if the weights and springs are not functioning correctly. It is not always easy to obtain the correct replacements, especially in the case of the springs, since these are different for almost every make of car and model. Tiny springs, costing two or three pence, with a long shelf life in the stores are not a popular stock item. Factory reconditioned distributors costing several pounds are readily available. One cannot stress too heavily the importance of fitting the correct weights and springs (or the correct reconditioned distributor). With a highly stressed sports car engine, a change in the automatic advance characteristics can endanger the life of the connecting rods, the bearings and the crankshaft. When a distributor analyser or a Crypton Motormaster is available, the functioning of the centrifugal advance mechanism can be checked against the maker's specification over the whole operative range.

Not all engines are fitted with vacuum units, but when fitted, a check should be made for the most common failing of this unit—a punctured diaphragm. This can be tested *in situ* in one of two ways:

(*a*) With the capillary tube from the manifold disconnected at the vacuum unit a thick-walled rubber tube is fitted over the union in the centre of the steel cover. A depression is then applied to the diaphragm by sucking with the mouth on the tube. The depression can be felt with the tip of the tongue and if the diaphragm is not air-tight the disappearance of the depression will also be felt by the tongue. A bad leak is easier to find by blowing through the diaphragm.

(*b*) With the capillary tube disconnected as before, the baseplate is rotated by hand against the action of the spring in the vacuum unit. The wetted forefinger is pressed against the end of the union in the steel cover. The baseplate is released and, with an air-tight diaphragm, it will be seen to rotate only part-way back towards its original position. After two or three seconds the finger is released, when the baseplate will be seen to return the rest of the movement to its original position. If the diaphragm is punctured no movement will occur when the finger is released.

If there remains any doubt after applying these tests the vacuum unit should be disconnected from the distributor and tested separately, either by the 'mouth and tongue' method or by the use of the vacuum pump in the distributor analyser. This latter test is of little value if the correct vacuum advance curve for the particular model is not known. If the maker's repair manual is not available, distributor advance data is often given in general garage repair manuals.

When the contacts are finally replaced and the voltage drop test has been carried out satisfactorily, it is a routine operation to adjust the gap to the correct setting. Remax Ltd make a special ring-type gauge that is very simple to use and quite dependable. When available, the tach-dwell meter should be used for the final fine adjustment, the gap being set to give the recommended dwell-angle rather than a specified gap.

Ignition timing

The final operation in tuning the ignition system is the timing operation itself. Experts differ on the value of the vacuum gauge for this purpose. After many years of tuning work in which he has used both methods, the writer now leans heavily in favour of the timing light.

The vacuum gauge method is simplicity itself. With the engine idling at about 500-600 r.p.m. and the vacuum gauge connected to a suitable tapping on the induction manifold as described in Chapter Two, the distributor locking clamp is loosened and the distributor head rotated gradually by very small amounts, retarding and advancing the ignition until the vacuum gauge shows the highest attainable reading. This should be the correct setting for maximum power, always supposing that the advance curve of the distributor is well matched to the engine. In the writer's experience the vacuum gauge does not always give the correct timing for the whole range of speeds and throttle openings. When the engine is a high compression competition engine he strongly recommends that the tuner who finds himself without a timing light should find the highest vacuum setting, then retard the ignition about one half-inch below this optimum setting. In general this seems to give a closer approach to the maker's timing and will remove the danger of excessive advance in any case.

The timing light is thoroughly dependable. The only error that can arise occurs when the timing marks on the front crankshaft pulley are almost inaccessible. The timing light should be connected across No. 1 cylinder ignition lead and any suitable earth. If a connection to the plug lead is difficult to make, a paper clip can be inserted in No. 1 terminal of the distributor cap, with one coil of the clip opened out, and the timing light lead clipped to this. Besides its normal use for setting the ignition timing at idle r.p.m. the timing light can also be used to check the maker's centrifugal advance curve. To check the centrifugal advance with nothing but an ignition timing light, it is necessary to mark out a band of degrees on the flywheel on the advance side of T.D.C. For clarity this can be done with white distemper or ceiling white, with the degree marks in black pencil. Once the flywheel diameter is known, every inch on the circumference of the flywheel is equal to X degrees, where

$$X = \frac{360}{\pi \times \text{diameter in inches}}$$

$$= \frac{115}{\text{diameter in inches}}$$

With the help of the tachometer we can now open the engine speed in steps of 500 r.p.m. and note the ignition timing at each speed. Since the engine is under no load the induction vacuum will remain high throughout the test and will not introduce any error on this count.

The vacuum advance curve can only be checked by means of the special distributor analyser described in Chapter Two, unless one chooses to make up a special rig with an electric motor and vacuum pump for this purpose.

Carburettors

The tuning procedure here has already been covered in Chapters Five and Six. Always remember to check that the throttles on all carburettors are synchronised and that they all can achieve the full-throttle position.

Road-test

This is a repeat of the tests carried out at the beginning of the test-tune schedule. One seldom finds that there is no measurable improvement. Occasionally the improvement is quite startling and the tuner returns to base with a warm glow in his heart.

This kit for Stage-2 Conversion of the Austin-Healey 'Sprite' and M.G. 'Midget' is made by Alexander Engineering Company of Haddenham, Bucks.

A racing option offered by Alexander is this cast alloy Cold Air Box for the 'Sprite'

PLATE 6

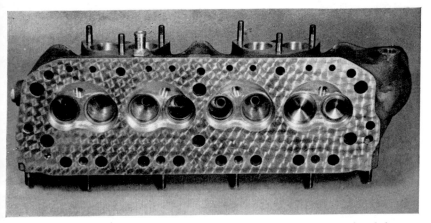

This aluminium alloy head for the 'B' Series B.M.C. engine, designed and developed by H.R.G. Engineering Co. and marketed by Alexander Engineering Co. Ltd. and V. W. Derrington Ltd. has four separate inlet ports on the opposite side from the exhaust ports. (The standard head has two siamesed inlet ports on the same side as the exhaust ports.) When used with the standard S.U. carburettor and a new inlet manifold power output is increased by 23 per cent. An 'A' Series alloy head is also available

A special rocker cover to suit the H.R.G. Series 'B' alloy head is also offered by Alexander Engineering Co. Ltd.

PLATE 5

Broken insulator nose. A clean break such as this is often caused by careless regapping. A pointed centre electrode with a side electrode that looks like new indicates an overheated condition

Flashover. The H.T. current will sometimes short across the surface of a dirty insulator. An excessive plug gap will increase this risk

(Photographs by AC Spark Plug Division, General Motors Corp.)

PLATE 4

ily plug. Usually caused by
cessive oil passing the piston
gs or down the inlet valve
ides. A hotter plug is a tem-
rary expedient.

Overheated plug. Burning or
blistering of the insulator nose,
with bad erosion of the electrodes, indicates an overheated
condition. Mixtures that are
too weak for full load operation,
or fuel that is too low in octane
value, could be the cause.
Check that the plug is of the
correct heat value

(Photographs by Champion Spark Plug Company, Toledo, Ohio)

PLATE 3

Normal plug appearance, indicating correct heat range and mixtures. Light brown or greyish tan deposits on insulator nose

Dirty plug. Dry fluffy carbon deposit indicates over-richness or plug running too cool for the particular engine design

(Photographs by Champion Spark Plug Company, Toledo, Ohio)

PLATE 2

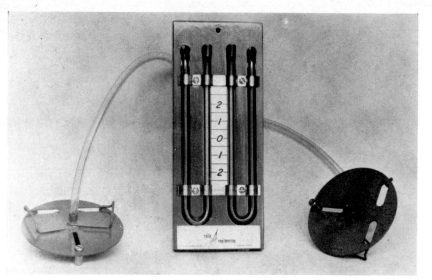

Multi-carburettor synchronising gauge marketed by Rally Engineering, 7140 Seward Ave., Niles 48, Illinois. An accurate method of balancing the idle air flow through two or more carburettors. Available for S.U., Solex and Zenith carburettors

The Rally Engineering device in use on a Sunbeam Alpine
(*Photograph by Paul Clymer, Racine, Wis.*)

PLATE 1

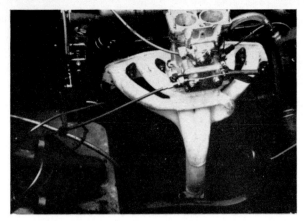

Twin-choke Weber carburettor kit for Triumph Herald by V. W. Derrington Ltd. of Kingston-on-Thames, England

S.A.H. Accessories of Leighton Buzzard, Beds., England, are specialists on the TR3 and TR4. Their racing engines, developing 135 b.h.p., are offered on an exchange basis. Note the design of the four-branch exhaust system

PLATE 7

Exhaust extractor devices appear in many forms. The Alextractor by Alexander Engineering Company of Haddenham, Bucks., is a well-engineered application of the ejector principle. At moderate to high road speeds the momentum of the air stream entering the annular shroud helps to extract the exhaust gases. Some pressure recovery occurs in the divergent section beyond the throat and under favourable conditions a sub-atmospheric pressure will exist at the throat of the extractor. When this occurs pressures throughout the length of the exhaust system will be correspondingly reduced

The life of the bearings in a supertuned engine can be preserved by a kit such as this. S.A.H. oil cooler kits are available for TR2, TR3 and TR4

PLATE 8

The Judson supercharger kit for the 'Sprite'. These vane-type superchargers are
marketed by Judson Research and Manufacturing Company of Conshohocken,
Pa., U.S.A., and by Performance Equipment Company of Birkenhead, England

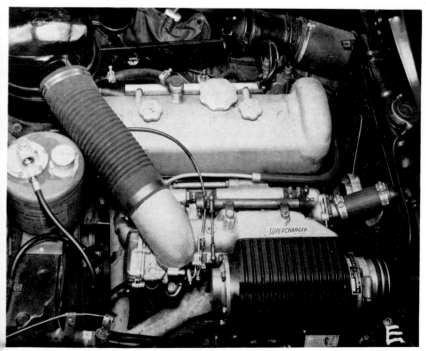

With the fitting of a Judson supercharger kit the Mercedes Benz 190SL is transformed into a 'tiger'

PLATE 9

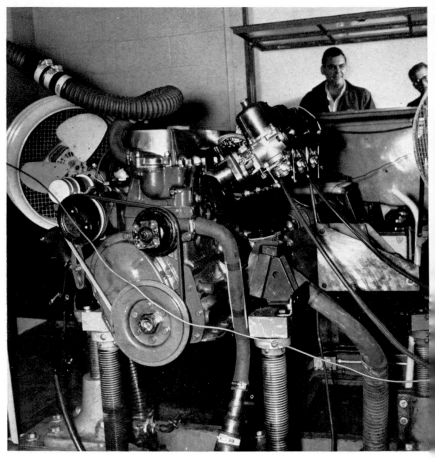

PLATES 10 to 15 give the case history of a successful 3-litre Austin-Healey. Hollywood Sports Cars of Los Angeles race-prepared two Austin-Healeys in 1961, the 2·6-litre owned by John Christy, editor of *Sports Car Graphic*, and the 3-litre owned by Chick Vandergriff of Hollywood Sports Cars. This latter car recorded nineteen straight wins out of twenty races. The 3-carburettor version of the 3-litre engine that gave 205 b.h.p. at 6000 r.p.m. is shown on the dynamometer. The final race version used a milder camshaft giving 195 b.h.p. and a wider range of useful torque

(*Sports Car Graphic* photograph)

PLATE 10

Chick Vandergriff (*left*) and engine builder Doane Spencer change carburettor needles during a series of dyno runs

(*Sports Car Graphic* photograph)

PLATE 11

The combustion chambers on the Vandergriff 3-litre Austin Healey were well rounded and polished. Note the complete removal of the valve guide bosses

(*Sports Car Graphic* photograph)

PLATE 12

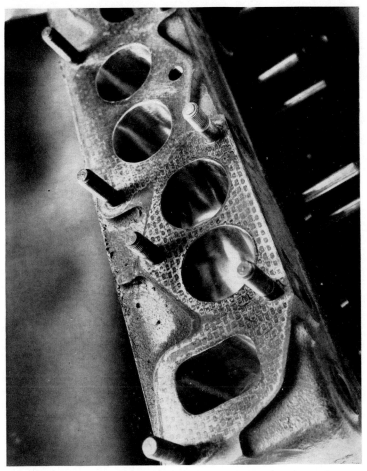

The ports in the Austin-Healey head were well polished, but were only opened out slightly. This helped to maintain a wide torque range, even when used with three HD6 S.U. carburettors. The final camshaft choice (designated HSC 'G' in Appendix 1) had a 274 degree duration on the inlet cams and 268 degrees on the exhaust. The lift was 0·405 inches

(*Sports Car Graphic* photograph)

PLATE 13

A polished and lightened rocker from the 3-litre Austin-Healey compared with the stock part
(*Sports Car Graphic* photograph)

PLATE 14

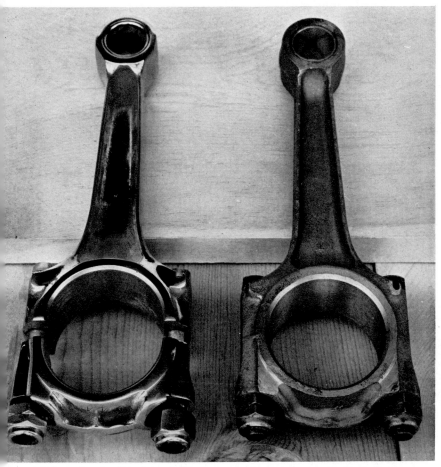

A lightened and polished Austin-Healey connecting rod compared with the stock product. Note that no substantial amounts of metal have been removed
(*Sports Car Graphic* photograph)

PLATE 15

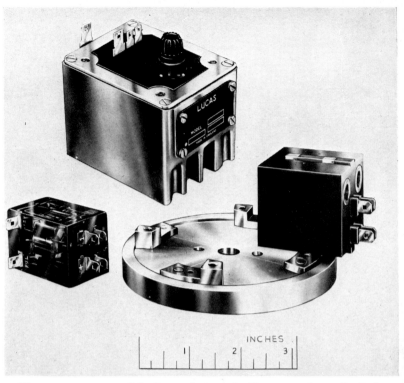

The component parts of the Lucas electronic ignition system for an 8-cylinder engine with fixed ignition timing

PLATE 16

CHAPTER EIGHT

Trouble-Shooting

ONLY the horse knows where the pain lies, but it is the rider who insists on talking to the vet. The motorist in a parallel situation will sometimes jump straight from the symptom to the cause and then to the cure. Quite unintentionally he can sometimes mislead the mechanic. The good mechanic always listens politely to the motorist's diagnosis, then methodically carries out the appropriate tests.

To the man who has learned the principles of ignition and car-buration purely by rule of thumb, hocus pocus or round the garage stove folk-lore, trouble-shooting is largely a game of chance with the poor car owner putting up the stake money. To the mechanic who really knows what goes on inside and around an engine it can lead to moments of pure triumph.

There must be method and sound logic behind any series of tests we may decide to carry out in our search for the culprit. The test procedure described in Chapter Seven does, in the majority of cases, provide sufficient evidence to enable us to trace and eradicate a fault in a particular engine. One should always carry out the full series of tests. This sometimes calls for a little self-discipline, but a good tuner has a reputation to maintain. Let us illustrate with a typical example of a half-completed schedule. The owner of the car has complained of bad starting and you have found the sparking plug gaps too wide and the insulator noses badly fouled. It is a great temptation at this point to hand the car back with a new set of plugs, especially if the owner is standing by and fuming at the delay. The very next morning the owner rings in, fuming even more strongly. The car won't start again. This time you discover that the fuel pump has a punctured diaphragm, but it is too late to regain the customer's confidence. You are labelled as a bungler. This then, is the reason for the insistence on the full test procedure. The owner may be a business executive with an important appoint-

SCE L   149

ment the following day; he may be a sports car competitor who is left on the start-line with an engine that fails to start. In either case he is a victim of your unjustifiable optimism.

But what happens if we have completed our schedule of tests and we have then proceded to carry out the tune and still nothing untoward has been discovered? Perhaps the customer has complained of an occasional misfire; of bad starting or of overheating and boiling on hills. Our series of tests has revealed no serious defects. What do we do next?

In a way we are in the same position as the Detective-Inspector who has made his routine cross-examination only to find that everybody has a cast-iron alibi. No Detective-Inspector worthy of an Agatha Christie or a Christiana Brand ever gave up at this point. Let's look for clues—and to do this we will need our vacuum gauge.

Fault-finding with the vacuum gauge

Much has been claimed for the vacuum gauge, especially by the firms that make them. Unfortunately, the vacuum gauge, used

TABLE 5

Typical Induction Vacuums for Engines in Good Tune

Make	Model	Idle vacuum reading inches of mercury
Aston Martin	DB2-4	15-16
Aston Martin	DB4	14-15
Austin Healey	2·6 litre Six	21-22
Austin Healey	3000 Mk. II	19-20
Austin Healey	Sprite Mk. II	19-20
Jaguar	XK 120, 140, 150	18-19
Lagonda	3 litre	15-16
Lotus	Coventry Climax 1100	19-21
M.G.	TC, TD	21
M.G.	TF 1250 and 1500	20-21
M.G.	MGA	19-20
Porsche	1600 Normal	17-18
Porsche	1600 Super	14-16
Porsche	1600 Carrera	13-14
Sunbeam	1·6 litre Alpine	19-20
Triumph	TR2	21-22
Triumph	TR3, TR3A and TR4	19-21

alone, is not selective enough. For example, a fluctuating reading can be caused by burned exhaust valves, sticking valves, a broken valve spring, a faulty sparking plug, a leak in the H.T. side of the ignition, etc. etc. Other tests are therefore required to pinpoint the true culprit or culprits. The use of the vacuum gauge in testing and tuning the normal engine has been described in Chapters Two and Seven. The idling vacuum of a typical sports car engine in good mechanical condition and good tune reads from 18 to 22 inches of mercury. Certain high performance engines with relatively large inlet valves, high lift cams or 'super-sports' ('three-quarter race' to the American) valve timings are most inefficient 'vacuum pumps' at idling speeds. Engines such as the Porsche 1600 Super only give an idling vacuum of 15 inches. It is therefore essential to know what vacuum reading to expect. The table on p. 150 gives typical vacuum readings from a range of sports car engines.

The behaviour of the vacuum gauge can sometimes give us a valuable clue to the source of a mysterious malfunction. The following notes should be of help.

Valve trouble

A periodic drop in the vacuum gauge reading at regular intervals by 2-4 inches is a fairly reliable sign of a burned valve, usually an exhaust valve. A further check with the compression tester, as described in Chapter Seven will serve to confirm this and show which cylinder is at fault.

A sticking valve will occasionally be revealed by a periodic drop in gauge reading, but in this case the downward kicks of the gauge needle are not at regular intervals. A squirt of penetrating oil or Redex into the appropriate carburettor will sometimes give a temporary cure. In this way the fault is immediately confirmed. Engines that are prone to valve-sticking will usually benefit from the addition of Redex to the fuel. Incorrect valve timing is indicated by a vacuum gauge reading that is an inch or two below normal, this reading, of course, being the highest obtainable after ignition timing and carburettor tuning are completed. The reading in such a case will be steady, but well below normal. The next step is a careful check of the actual valve timing and the necessary correction to the position of the timing gears or sprockets.

Induction leaks

A low vacuum reading at idle could be an indication of an air leak in the induction system, but a more positive indication is given when the engine is turned over on the starter against a fully closed throttle. When the car is fitted with a combination ignition and starter switch it will be necessary to disconnect the H.T. lead from coil to distributor, or if convenient, to remove the distributor cap. For this 'motoring' test it is absolutely essential that the throttle plate or plates are completely closed. The throttle-stop screws should be turned back until they no longer bear on the throttle-plate abutments. In this position the throttle plates will almost completely seal off the induction system from the atmosphere; the degree of seal depending upon the fit of the plates in the bores and the amount of wear in the throttle spindles and their bushes.

One cannot state precisely what vacuum to expect when the engine is turned over by the starter motor. Since all leaks present are of fixed area, the higher the cranking speed the higher will be the vacuum pulled by the engine. Thus a tired battery on a recently rebored engine would give a low reading and a well-charged battery on a free-running engine would give a high reading —all other factors being equal. One therefore requires experience to interpret the result. Only the extremes can be treated as certainties. If the vacuum reading is as high as or higher than that recorded when the engine is idling one can feel certain that no leaks of any consequence can exist in the induction system. At the opposite end of the scale a reading of 6 inches or less is a sure sign of a serious leak. A vacuum reading that is 5-8 inches below the normal idling figure would be quite normal on an old engine.

The most common source of leaks in the modern induction system is the carburettor to induction manifold joint. Many sports car engines are fitted with heat-insulating packings at this point. Ebonite makes a durable packing but one should always anticipate trouble when the standard packing is of asbestos or other plaster, with paper backing faces. The compressive strength of this material is not high and it is not uncommon for it to contain small cracks that are not readily seen by the naked eye. Any joint or packing, or any point between the carburettor and cylinder head where a leak is suspected, should be liberally coated with oil when the engine is idling. If any oil is drawn through the leak it will be revealed by a puff of blue smoke at the exhaust. A check should also be made on the tightness of external connections between manifolds. On the earlier Aston Martins, for example, there is an inconspicuous rubber grommet at the junction between the front and rear induction manifolds. This is sometimes a source of air leakage. Finally one should not forget the vacuum gauge tapping. If in any doubt, smear some gasket jointing compound on the threads of the adaptor and refit.

Broken or defective piston rings

The use of the compression gauge to check the condition of the rings and bores has been described in Chapter Seven. Occasionally the indications that the rings are not sealing too well are not very clear-cut and confirmation by vacuum gauge is then called for. A quick check can be made as follows: Open the throttle until the engine speed rises to about 4000 r.p.m., then close the throttle quickly. If the rings and bores are in good condition, while the engine speed is falling the vacuum gauge will rise to a higher reading, momentarily reaching a reading 3 to 7 inches higher than the normal idle vacuum reading, before finally falling to the normal idle value. A failure to register this rise or even a showing of one of only 1 to 2 inches is an indication of poor ring seal. To check the individual cylinders, the engine should be run at about 1000 r.p.m. and each plug shorted in turn. There will be a drop in vacuum reading as each plug is shorted, but a cylinder showing a larger drop than the average is almost certainly the culprit. It is assumed at this stage that the plugs are not suspect, but a check on this point by switching plugs between a good cylinder and the suspect cylinder will prevent any mistake on this score. Finally a repeat of the dry and wet compression readings should be taken to make certain that the fault lies in the pistons and bores and not in a leaky valve.

Defective exhaust system

Abnormally high exhaust back-pressure is not a common failing in sports cars since a straight-through silencer is normally fitted. Occasionally however a case arises in which a front pipe or a tail pipe has been accidentally flattened during the passage over rougher sections of a rally course. Hardened rallyists are apt to forget to report such occurrences, especially when they occur in the early part of a long and arduous rally. An outstanding case that had already baffled an experienced tuner provided a moment of triumph to a member of the writer's staff when a Wolseley 14 with a top speed of 45 m.p.h. and acceleration to match was brought in to the tuning bay and was found to have no fault in carburation or ignition or compression.

The following test for a choked exhaust system was carried out: the engine is raced up to about 4000 r.p.m., then the throttle is closed quickly, as in the piston ring seal test. With a choked exhaust system the time taken for the vacuum gauge to fall back to a normal idle reading—after the customary rise—is of excessive duration. Experience with normal exhaust systems soon establishes the duration norm. In the case of the Wolseley the duration was very long indeed and the tester felt no compunction, after checking that no obvious restrictions existed in the front pipe and tail pipe, in cutting open the silencer for examination. Inside he found one of the most complex tortuous baffle systems ever devised by an amateur. A new standard silencer restored the power.

Much time has been devoted to the use of the vacuum gauge as an instrument of diagnosis—perhaps too much. Our only excuse is that the instrument is inexpensive and the time taken in carrying out the tests, once the gauge is connected to the manifold, is very short.

A short catalogue of troubles

In Chapter Seven we considered the general case of testing and tuning an engine which is not in good tune, but is otherwise able to operate satisfactorily on the road. Trouble-shooting, however, embraces the wider field of engines that fail to start, engines with occasional misfires, engines that run too hot, engines that suffer from all the ills known to Man and some known only to the Devil. The whole subject is a big one and the writer knows that he will fail to mention some of the more uncommon troubles, but the common major and minor misdemeanours of the petrol engine are described and the cures listed in the next few pages.

1. *Engine will not turn over, or turns over very slowly*

(*a*) Remember—this is not always the fault of the battery. Switch the headlights on. If they only give a dim light check the condition of the battery and its connections. The procedure is laid down in Test No. 5 in Chapter Seven.

(b) If the lights are bright, press the starter button and notice the effect on the headlight beam. If the lights stay bright and are not apparently robbed of current there must be a fault in the motor circuit. The connections to the starter and starter solenoid should be checked for tightness and cleanliness. If there is no fault to be found here, the motor should be removed, the brushes examined and the windings checked for continuity. From this stage the work on the motor usually becomes the concern of an auto-electrical specialist.

(c) If the lights go very dim as the starter button is pressed and the battery has been shown to be in good condition the starter drive pinion should be examined. Occasionally a pinion will jam on the flywheel teeth, especially if the teeth are worn. Sometimes the armature can be rotated by hand without removing the motor from the clutch housing. If the armature will not turn it should be removed for examination. Sometimes, as a result of excessive end-play in the bearings, the motor is free to turn when rotated by hand, but locks with excessive friction when under load, i.e. when driving the Bendix pinion into mesh. If, however, the battery is in good condition and the starter is not at fault, there is only one place left to look—the engine. Oil and water should be checked and if the engine is too stiff to turn over by hand one should proceed to check for a seizure or breakage of a rotating or reciprocating component.

(d) If the whine of the starter motor can be heard when the starter button is pressed, but the engine does not turn over, the Bendix pinion should be examined. A sticking pinion is usually caused by dirt, but excessive wear on the flywheel teeth can sometimes put an abnormal side load on the pinion shaft, especially if the pinion has been occasionally failing to engage with the flywheel teeth for a certain period before finally jamming. This final jamming is caused by the armature shaft eventually bending under the ill-treatment.

2. *Engine turns over at normal speed, but fails to start*

(a) Check strength of spark at plugs by disconnecting one or more leads and holding about $\frac{3}{16}$ inch away from the metal of the head. If the spark is feeble or non-existent the whole ignition system should be checked out as described in Chapter Seven, starting at the plugs and going through the coil, condenser contacts, plug leads, etc.

(b) If the spark is good, the induction system should be primed with fuel, by jiggling the accelerator pump link, by flooding the carburettor bowl or even by squirting some petrol into the carburettor intake. If the engine then starts, runs for several seconds, then stops, we have proved that the trouble lies in the fuel supply. A start should be made by disconnecting the pipe between the fuel pump and the carburettor, at either end, and observing the fuel flow when the pump is operating. In the case of the A.C. pump it will of course be necessary to turn the engine over by starter. If the fuel pump appears to function but at a low pressure and reduced delivery rate, look for air bubbles in the fuel. This is usually caused by a leaking fuel line between the pump and the tank.

Fuel pumps and their more common troubles are discussed later in the chapter.

(c) If the fuel pump is delivering an adequate supply of fuel to the carburettor the carburettor should be examined. A blockage in the idling system, dirt in the float chamber inlet filter or a bad air leak at the carburettor/induction pipe gasket are possible defects.

(d) If the carburettor is clean and in good working order, a check should be made for an air leak in the induction system in the manner described earlier in this chapter.

(e) If everything checked so far has been sound, examine the appearance of the plugs for water. Drops of water on the insulator nose or electrodes can indicate an internal leak, usually from a faulty head gasket seal. If the leak is bad a spray of water droplets will be seen issuing from an open plug socket when the engine is turned over on the starter.

3. *Engine runs but misfires*

As soon as we are presented with a running engine, however badly it may run, our problems are simplified. Dead engines don't talk. Testing can proceed initially along the lines of the normal test schedule laid down in Chapter Seven.

(a) If the misfire occurs when the engine is not under load the location of the trouble should be straightforward, since the normal test instruments will soon pinpoint the defective component.

(*b*) A misfire that only occurs under load cannot be reproduced in the garage, unless of course the garage is equipped with a chassis dynamometer. If no defects or maladjustments are revealed by the test schedule of Chapter Seven then our final resort is to take to the road and this to-day, with crowded roads—sometimes even no place to draw off the road to make tuning adjustments—can discourage the hardiest tester.

4. *Misfire or loss of power at high speeds*

(*a*) *Check that the air cleaners are not dirty.* Run with them fitted, then without. Clean or replace as thought necessary. A badly choked air cleaner will give an over-rich mixture, especially at high air flows.

(*b*) *Check for pre-ignition.* Drive the car hard up to the point where the misfiring or loss of power occurs, then cut the ignition, drop out of gear and coast to a standstill. Remove the plugs and check for signs of overheating. The insulator nose will be clean and white, perhaps slightly speckled. Sometimes the nose of the insulator is eroded. A typical case is shown in Plate 3. When pre-ignition has been going on for some time, particularly in conditions when the throttle can be held down for long periods, as on a modern express-way, the end of the central electrode can be completely eroded away together with a large part of the insulator material and the outer electrode too. The fact that the gap has become impossibly large has no influence on the abnormal combustion that is now taking place. Ignition is no longer controlled by the normal ignition system but occurs in a completely uncontrolled manner from some overheated surface, usually the central electrode of the overheated plug.

If pre-ignition is confirmed by this examination, there are two possible causes. First, the plugs in use may be of the wrong heat value for the particular engine. Second, a fault in carburation may be causing the carburettor to feed too weak a mixture at high speed. Overweakness as a cause of pre-ignition becomes more prevalent the more highly supertuned the engine. An engine with a low compression ratio and a low power output would be a most unlikely subject for this type of pre-ignition.

(c) *Check for loss of power from incorrect carburation or fuel starvation.*
The use of a portable exhaust gas analyser helps to clear up any
doubt here. In the case of S.U. carburettors the needle type
should be checked against S.U. List No. AUC 9631. When fixed
choke carburettors are fitted the sizes of the jets should be checked
against the maker's specification. If any major tuning modifications
have been carried out on the engine no such basic data will be
available and the only method left to us, if an exhaust gas analyser
is not available, is to experiment with jet changes. Increase the
main jet on an old type Zenith; decrease the air correction jet on
the new Zenith, the Solex and the Weber. This will probably give
an increase in power. If, however, the loss of power is even more
pronounced, the indications are that the mixture strength is far
too rich at this part of the power curve. Finally, we have the case,
much more rare, of an increase in jet size making no change at all.
Here we can assume that the change in jet size has had no influence
on the mixture strength since the fuel is restricted before it gets to
the jet. If the engine has been extensively modified it is possible
that the increased power output has taken the fuel consumption
beyond the normal delivery rate of the standard fuel pump. The
writer can remember such a case when a standard car was fitted
with a fairly large supercharger. The owner had tried every type
of plug he could find, but the trouble lay in the inadequate fuel
supply to the carburettor.

(d) *Check for inadequate spark at high speed.* With coil ignition the
voltage drops as the engine speed increases (see Fig. 16 in Chapter
Three). With all the ignition components in good condition and
the ignition system in a good state of tune there should be a high
enough voltage produced in the secondary windings for an effective
spark to occur every working cycle at the top end of the speed
range. Any defect in the system, however, will usually show up at
high speed under load. If the appropriate tests of Chapter Seven,
i.e. Nos. 3, 4, 5, 6, 7 and 8 have shown no defects in plugs, coil,
condenser, battery, etc and the misfire at speed persists, one must
look deeper into the ignition system. To be absolutely fair to the
ignition system when one has reached this 'leave-no-stone-un-
turned' stage it is worthwhile to carry out a high speed test with the
ignition system specially tuned to favour top-end spark. The plug
gaps should be reset 2 or 3 thousandths of an inch below standard

and the contacts should be set 1 to 2 thousandths fine to give a longer dwell angle. If this setting, giving slightly increased secondary voltage at high speed, does not cure the high speed miss, we should start to examine the ignition system with the proverbial fine tooth comb. Here are four possible hidden defects.

(i) *Poor distributor earth.* If the electrical contact between the distributor body and the metal of the engine is poor, the added resistance of this poor earth will reduce the voltage across the windings and, in turn, this will reduce the secondary voltage. Several cases of persistent high-speed misfire, in the writer's experience, have been cured by fitting a braided copper earth strap between the distributor casing and a convenient bolt or stud on the engine.

(ii) *Faulty wiring or connections in the primary circuit.* The continuity of this circuit is easily checked, but a component that is sometimes neglected is the ignition switch. Excessive voltage drop at this point can rob the coil of those vital volts that are needed at high speed under load.

(iii) *Cross-fire between H.T. leads.* When plug leads are fitted in close proximity in a harness, or as in certain older cars, are threaded tightly through a fibre tube, it is possible for an electrical charge to be induced in a neighbouring lead by the rapid voltage build-up in the correct lead. Although the induced voltage will be lower, the plug fed by the 'robber lead' has the advantage of a lower cylinder pressure and with favourable conditions it is possible for a spark to occur in the wrong cylinder. If this occurs during the induction stroke or early part of the compression stroke we have the phenomenon known as 'cross-fire'. That cure is obvious— separate the leads. If it appears that cross-fire could be occurring with the standard plug lead layout, it is a simple matter to make a temporary lash-up with the leads arranged well apart. For example the leads can be attached to flat strips of wood by means of insulation tape, giving a space of about half an inch between each lead. If this effects a cure of the misfire when road-tested a neater installation can then be devised. If it has no effect at all, the original layout can be retained. On most modern engines great care has already been taken in the layout of the plug leads to avoid this danger.

(iv) *Tracking at the rotor arm or across the distributor cap or breakdown of the plug lead insulation.* It is to be expected that an intelligent tuner will by now have wiped clean the plug leads and removed all trace of road film from the distributor cap. Nevertheless it is still possible for a spotlessly clean cap, with no cracks in its surface that are visible to the naked eye, to be breaking down in its insulation when subjected to high voltages. Certain proprietary garage tuning kits, such as the Crypton BX Series Analyser, supply a H.T. Search Tester. Essentially this is a single electrode fed with a high frequency high voltage continuous 'shower' of sparks. By holding the probe by the insulated handle a search can be made by passing the tip of the electrode over the surface of the rotor arm, the distributor cap and along the length of each ignition lead—a search for signs of 'tracking'. Without such a probe, one can only use one's judgement of the condition and age of the various components in the H.T. circuit. The cost of a new rotor arm and a new distributor cap is really quite modest in any case.

(*e*) *Check for inadequate fuel supply.* The fuel pump will have been checked before this stage, but the following points should be investigated.

   (i) dirty carburettor filter.

  (ii) dirty fuel pump filter.

 (iii) air leak between tank and pump. This will be revealed by air bubbles in the delivery line. One or two small bubbles that stay static in a bend in a plastic fuel feed pipe can be ignored.

 (iv) flattened or defective fuel pipe.

  (v) blocked air vent to tank. The writer remembers two cases of burned exhaust valves that were traced to over-weak mixtures under high speed load caused by blocked vent-holes in the tank filler cap.

 (vi) *fuel pump normal output no longer adequate.* Occasionally the increased demand of a tuned engine exceeds the normal delivery of the standard pump. Sometimes on a 'special' the amateur designer has underestimated the fuel requirements of his engine. The cure is obvious; the installation of a larger pump or the addition of a second pump in tandem.

5. *Misfire or loss of power at low and medium speeds*

Several of the high-speed defects listed under 4 could, in certain cases, show up at lower speeds. Such defects as pre-igniting plugs will show up at medium speeds when climbing a gradient. Ignition defects, if serious enough, will occur at lower speeds and will prevent the attainment of higher speeds. Usually, however, such defects are serious enough to have been revealed by the normal garage testing procedure of Chapter Seven. The following defects, however, are more prone to occur at low speed and low load than at high.

(a) *Overcool plugs*. If the misfire occurs after idling in traffic, one or more plugs will be found to be wetted with petrol. If the trouble persists a change to a hotter running plug should be made. On certain high performance cars this could lead to pre-ignition under maximum power high speed condtions. For such cases a reasonable compromise is not possible and the driver should keep the engine running at a fast idle during traffic stops, with an occasional 'blip' on the throttle to clear the plugs. Sometimes a change to platinum point plugs will effect a cure, since these plugs have a wider operating temperature range.

(b) *Faulty carburation*. If the mixture appears to be correct at high speed, but not in the medium speed range, changes in jet size should be made, but only one jet (i.e. one per carburettor) should be changed at a time. In the case of the Solex, Weber and the new Zenith carburettors it is the main jet that has the greater influence on medium speed metering. An increase in jet size should be tried first, then a decrease. In the case of the older Zenith carburettors the appropriate jet is the compensating jet. If a marked improvement in performance is given by a change in jet size, the final tuning of the carburettors can be made by a careful road-testing of small changes in the complementary jets too, i.e. the air correction jets in the Solex-type carburettors and the main jet in the older Zeniths.

(c) *Water in the fuel*. This is not a common occurrence, but the presence of water in the fuel will usually be found by looking inside the float bowl. If the bowl is easy to remove tip its contents into a funnel lined with chamois leather. The telltale blobs of water will soon be spotted on top of the chamois. Drainage of the tank is the only permanent cure. Resist the temptation to rush off in a hot temper to your local filling station. The petrol suppliers carry out periodic checks for the presence of water in underground tanks.

Moreover the filling station proprietor takes great care to see that water does not seep into his tanks. The chances of water entering the car's petrol tank are far higher. If the trouble persists the fit of the filler cap on the tank spout should be carefully checked. Occasionally a filler spout is so situated that rain can get into the tank. Where the filler is recessed into the bodywork the blockage of the water drain from the recessed compartment can cause a back-up of rain-water around the filler spout in heavy rain storms.

(d) *Oily plugs*. An engine that consumes excessive quantities of lubricating oil will sometimes put a plug out of action by the deposition of oil across the points. The excessive oil consumption, however, will already have been shown, in no uncertain manner, by the clouds of blue smoke from the tail pipe, especially when opening up the throttle at the bottom of a long hill. The common cure for this major mechanical defect is well known. Only one note of warning is necessary. Obsessed as the typical repair garage is with the problems of piston ring and bore wear, the possibility of oil entering the combustion chamber down the inlet valve guides is sometimes overlooked. The appearance of an oily plug is shown in Plate 3 (see pp. 148-149).

(e) *Water on the plugs*. An internal water leak from a faulty head gasket, or a corroded or frost-cracked cylinder head, can sometimes cause a plug to misfire. When the leak is serious the engine will almost certainly fail to start. If the leak is minor, the engine will run, but with an occasional misfire occurring when a drop of water splashes on the plug tip.

The indication of water on the plug electrodes will only be found when the engine is cold. Confirmation can also be obtained by a check on water loss from the cooling system. Leakage of water into the combustion chamber is often accompanied by a reverse flow during the firing stroke of combustion gases into the cooling system. With the engine at running temperature, the sight of gas bubbles rising to the surface of the water in the header tank, can be taken as further confirmation of an internal water leak.

(f) *Sticking valves*. Exhaust valves seldom stick, but when they do the condition soon leads to the destruction of the valve seat. A sticking exhaust valve is usually indicated by backfiring in the exhaust system.

Intermittent sticking of an inlet valve is a more common trouble and can be caused by:

- (i) too much or too little clearance in the valve guides,
- (ii) excessive gum deposition from the fuel used,
- (iii) carbon deposits from exhaust blow-back during the valve overlap period. Engines with large overlap valve timings are prone to this trouble if run for long periods on small throttle openings.

If the valve sticking can be produced at idle, the vacuum gauge can be used to verify the fault. A burned or leaking exhaust valve or inlet valve will show a *regular* periodic drop in the vacuum gauge reading. A sticking valve will show an *irregular* occasional drop of 2-4 inches on the vacuum gauge. If three or four squirts of Redex or other upper-cylinder lubricant into the carburettor intakes brings about a temporary improvement the trouble is known to be situated in one of the inlet valves. Even if the trouble only occurs when under load a good dosing with Redex into the carburettor intakes will effect a temporary cure. For a more permanent cure one must carry out a top overhaul with a careful check on valve guide clearances.

6. *Overheating*

Boiling of the coolant in the radiator under all but extreme 'Arizona summer' conditions indicates one of two things; either an abnormal heat input to the cooling system or an inefficient cooling system.

At this stage of our investigation there should be no doubt left that the ignition timing is correct or that the mixture strength provided by the carburettors is normal. A late ignition timing—too retarded—gives late burning of the mixture and increases the heat-load to the cooling system. A mixture that is too weak burns more slowly than a normal mixture and is still burning when the exhaust valve opens. This again puts an abnormal heat input into the cooling system. Pre-ignition is another combustion phenomenon that increases the heat rejection to the cooling system. In its more violent forms pre-ignition is characterised by a very harsh combustion knock. When less severe the combustion 'roughness' could easily be overlooked, especially if the tuner is not familiar with the sound of the engine under observation.

More causes of overheating are listed below:

(*a*) slipping fan belt,
(*b*) defective thermostat,
(*c*) defective radiator cap,
(*d*) internal water leak,
(*e*) slipping clutch,
(*f*) incorrect valve timing.

FUEL PUMP TROUBLES

Nothing was said of the internal construction of fuel pumps in the earlier chapters. Tests for output and delivery pressure were described in Chapter Seven, but a short discussion of the design of the two well-known types of British fuel pump and the usual defects encountered in service appears appropriate here.

The A.C. fuel pump

Fig. 53 shows the working principle of all mechanical type pumps, such as the A.C. A downward pull on the rod attached to

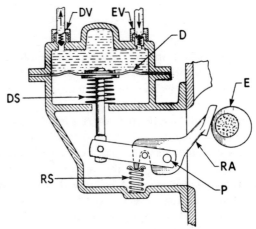

Fig. 53. Basic A.C. pump

the diaphragm *D* compresses the spring *DS* and creates a depression in the upper chamber. Fuel under atmospheric pressure then flows through the inlet valve *EV* until the pressures are equalised. As the deflecting force of the cam lobe *E* is removed by rotation of the cam, pressure can build up in the upper fuel chamber under the action of the spring *DS*. If no fuel is demanded by the carburettor float chamber the diaphragm will remain in this position, maintaining a steady pressure in the upper chamber. The primary cam lever *RA* will be free to move in and out under cam action, leaving the secondary lever and the diaphragm undisturbed. Any usage of fuel, however, will raise the position of the diaphragm and on the next downstroke of the primary cam lever *RA* the charged position of the diaphragm will be restored.

A.C. pump troubles

1. *Low pressure or reduced delivery rate*
    (*a*) weak spring.
    (*b*) stretched or punctured diaphragm. Since a punctured diaphragm can usually let petrol into the crankcase the oil should be changed when this occurs.
    (*c*) air leak on the inlet side of the pump.
2. *Reduced delivery rate*
    (*a*) restricted delivery line, from damaged flex hose or a blocked pipe.
    (*b*) choked fuel filter.
3. *Fuel pressure too high*
    (*a*) omission of gasket between pump and crankcase joint faces.
    (*b*) erroneous fitting of incorrect model of replacement pump.

The S.U. pump

The action of the S.U. pump can be followed from Fig. 54. The magnetic attraction of the solenoid winding *SW* pulls the plunger *P* downwards until at the bottom of the stroke the toggle mechanism *T* flips open the contacts *C* which breaks the current supplied to the solenoid *SW*. The plunger then returns under the action of spring *DS*. The function of the diaphragm, inlet and delivery valves is identical to that of the A.C. pump.

S.U. pumps are made in three models: L, HP and LCS. The older design, the L type, is designed for mounting near the engine at about carburettor level. The maximum suction lift is 48 inches and the maximum delivery head 24 inches. The HP pump is distinguished by its longer body. This type is usually mounted near the fuel tank and is designed to give a suction lift of 33 inches

SCE M

and a delivery head of 48 inches. Maximum output is 10 gallons per hour against the 8 gallons per hour of the L type. LCS are again designed for tank mounting and have a higher output still, 12·5 gallons per hour.

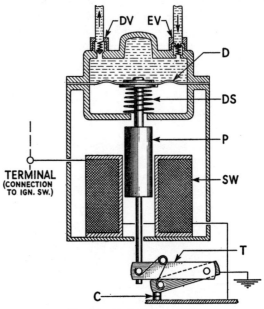

FIG. 54.   Basic S.U. pump

The component parts of the HP pump are shown in Fig. 55. There are three main assemblies, the body, the magnet assembly and the contact breaker. The body, 8 in Fig. 55, contains the outlet union, 1, which holds the delivery valve disc, 4, and the delivery cage, 5, in position. Below the delivery cage is the suction valve disc, 7, which is held on a seating machined in the body casting by a light spring, 31. The pump inlet union enters the body at an angle and is just visible in the figure (see 29). The pumping chamber is a shallow cavity to the left of the diaphragm, 9, which is connected to the space between the two valves by communicating holes. The outer edge of the diaphragm is clamped between the magnet housing, 27, and the body, 8. The centre of the diaphragm is held between the retaining plate, 11, and the steel armature, 15. A bronze push rod, 16, is attached to the diaphragm and screwed through the centre of the armature. This push rod passes through the magnet core to act on the contact breaker located at the left-hand end of the pump. Between the magnet housing and the armature are eleven spherical-edged brass rollers, 10. These act as

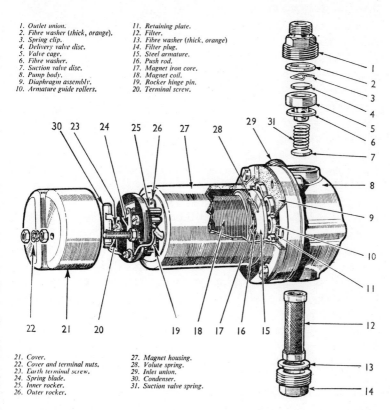

1. Outlet union.
2. Fibre washer (thick, orange).
3. Spring clip.
4. Delivery valve disc.
5. Valve cage.
6. Fibre washer.
7. Suction valve disc.
8. Pump body.
9. Diaphragm assembly.
10. Armature guide rollers.
11. Retaining plate.
12. Filter.
13. Fibre washer (thick, orange)
14. Filter plug.
15. Steel armature.
16. Push rod.
17. Magnet iron core.
18. Magnet coil.
19. Rocker hinge pin.
20. Terminal screw.
21. Cover.
22. Cover and terminal nuts.
23. Earth terminal screw.
24. Spring blade.
25. Inner rocker.
26. Outer rocker.
27. Magnet housing.
28. Volute spring.
29. Inlet union.
30. Condenser.
31. Suction valve spring.

FIG. 55.   The S.U. pump, type HP

guides of low frictional drag to locate the armature centrally within the magnet. At the end of the magnet housing is a small Bakelite moulding, the contact breaker. Below the Bakelite mould-ing are two rockers, an inner, 25, and an outer, 26. The rockers are spring-loaded to give an 'over-centre' or throwover action. The outer rocker, 26, is fitted with a tungsten point which makes contact with a second tungsten point on the spring blade, 24. This spring blade is attached electrically to one end of the coil. The other end of the coil is connected to the terminal, 20. A short length of wire connects the outer rocker, 26, with the other terminal, 23.

S.U. pump troubles

When an S.U. pump is suspected, the first test should be to disconnect the outlet pipe to the carburettors and check the delivery rate. If this is up to standard and the pressure is also found to be within the manufacturer's limits, as laid down in Chapter Seven, then suspicion should be redirected to the carburettors and the fuel pipes. If the pump fails to work the electrical lead should be disconnected and a test-light used to check that current can flow to this point. *Never check for current flow by sparking the lead to earth.* If the pump is leaking the resulting explosion could be disastrous. If it is shown that current is flowing to this point the Bakelite cover should be removed from the contact breaker assembly to check that the contacts are touching. If the contacts are touching, it is probably dirt between them that is preventing the flow of current. This can be removed by placing a thin piece of cardboard between the points, pressing them together and pulling the card backwards and forwards two or three times. Never use a file or emery cloth.

If the points are not making contact, check that the tips of the inner rocker, 25 in Fig. 55, are in contact with the magnet housing. If they are not in contact it is probable that the armature cannot complete its normal stroke. It will be necessary to dismantle the pump at this stage to find out the reason for this. Excessive wear or an accumulation of dirt inside the pumping chamber from a punctured filter are possibilities.

At this stage, if the reader is not experienced with this type of pump it is perhaps a wise plan to exchange the pump for a factory reconditioned unit. There are several tricky adjustments to be made when setting up the S.U. pump and several pitfalls exist in the technique of re-assembly. For the time involved in replacing points and diaphragm without correct assembly tools, the replacement pump represents good value for money.

If the pump is noisy and works at an excessively high pumping rate it is highly probable that the suction pipe is drawing in air. If the level of fuel in the tank is found to be high enough a quick check for the existence of an air leak on the suction side can be made by immersing the delivery pipe in a jar of fuel and looking for the presence of air bubbles as the pump delivers fuel into the jar.

If the pump fails to deliver fuel but makes a beating noise the valves should be removed and examined. The most common cause of faulty valve seating is the presence of dirt in the fuel.

Simple pump maintenance is often neglected, even by the enthusiasts. The pump filter should be cleaned regularly, at least every 3000 miles. At the same time the tightness of the electrical connections should be checked. These seem to have a habit of slackening off.

Don't shoot the trouble-shooter—he is doing his best

Variety is the essence of the tuner's life and some of the troubles he encounters occur at such infrequent intervals that he cannot be expected to remember everything. Here is an example of the infrequent type of trouble that the customer sometimes expects to be solved immediately. The owner of a high-performance specialist sports car complained of a complete failure of ignition on certain occasions. The car would behave perfectly for several days, then suddenly the engine would cut out completely and refuse to restart. After a wait of ten to fifteen minutes, however, a restart could be made and the engine would run perfectly. The tuner forgot to ask, and the owner failed to observe, that the trouble only occurred on wet days and, even then, only when there was a cross-wind blowing the rain on the right-hand side of the car. Much time and money was spent in trying to locate this fault, a fault that could not be reproduced in the garage. Finally an observant mechanic noticed that a fresh film of road-dirt had appeared on the side of the distributor body since his last examination. Only a week had passed since he last wiped down the distributor. From this he deduced that dirty rain water was passing through a gap in the left-hand wheel arch and spraying on to the distributor. The gap was found and sealed and the trouble never recurred.

Experienced tuners have hundreds of tales like this to tell and the moral is always the same: Use your eyes, then use your head.

CHAPTER NINE

Combustion Efficiency

Supertuning

This is the first of five chapters covering the many aspects of supertuning, or as some prefer to call it, engine modification.

Supertuning is simply the practice of increasing the power output of a standard engine by mechanical modification. The tuning that has concerned us in the earlier chapters is the application of certain techniques to ensure that a standard engine is giving its peak performance. When a standard engine has been supertuned this peak will be lifted to a new level, but the tuning techniques of Chapter Seven will still be required to maintain the supertuned engine in perfect tune.

A post-war development of this near-ancient art has been the spread of 'Stage Tuning', that is, factory approved stages of power enhancement, together with full instructions and a spares service to help the owner mechanic who wishes to supertune his own car. The M.G. Car Company were probably the first in the field of stage tuning with the introduction of five Tuning Stages for the XPAG series engine in the TC Midget. To-day stage tuning has developed into a kind of extra-mural development department for all reputable manufacturers of sports cars. The whole process tends to be synergetic, the manufacturer helping the tuners and the more advanced free-lance tuners feeding back new knowledge, new tricks, to the factory.

The great advantage of manufacturers' stage tuning is that the car owner is assured that the degree of supertune is not excessive. Each of the suggested (not recommended!) stages has been thoroughly bench-tested and road-tested by the maker's development department and is known to preserve what we may call 'a reasonable standard of reliability'. No one can ever give a guarantee of absolute reliability in the field of supertuning and the competition driver must always approach the business philosophically.

THE BASIC PRINCIPLES OF SUPERTUNING

Methods of power-boosting can in general be divided into two categories:

(a) those that improve combustion efficiency
(b) those that improve volumetric efficiency.

This chapter covers the first of these, while Chapter Ten discusses Volumetric Efficiency.

Combustion efficiency

The amount of useful work we can extract from the combustion of a pound of fuel in an internal combustion engine depends upon several factors, but the major influence is that of compression ratio. It is customary in textbooks of thermodynamics to apply a theoretical concept of an 'Air Standard Efficiency' (sometimes called the Otto Cycle Efficiency in the case of the petrol engine). In this concept all the fuel is considered to be burned instantaneously at top dead centre and the specific heats of the gases in the cylinder are considered to be equal to those for air.

The Air Standard efficiency is given by the following formula:

$$E = 1 - \frac{1}{R^{K-1}} \qquad (1)$$

where R = the compression ratio

and K = the coefficient of adiabatic expansion, which is equal to the specific heat of air when heated at constant pressure divided by the specific heat for air when heated at constant volume
          = 1·4 approx.

A plot of E against R is given in Fig. 56. No practical engine ever achieves the efficiency of the theoretical Air Standard. Fuel cannot all be burned at the moment of maximum compression and fuel burned late in the stroke contributes less useful work than fuel burned at T.D.C. There are practical limits to the rate at which we can permit fuel to be burned inside the cylinder, the most important being the mechanical limitations set by the strength of the cylinder head. Almost as important is the undesirable increase in combustion noise that usually occurs when combustion pressure rises at a greater rate than 40 lb. per sq. in. per degree of crank rotation.

The value of 1·4 for K is based on adiabatic conditions, i.e. no gain or loss of heat to external sources during compression or expansion. There must always be some loss of heat through the walls

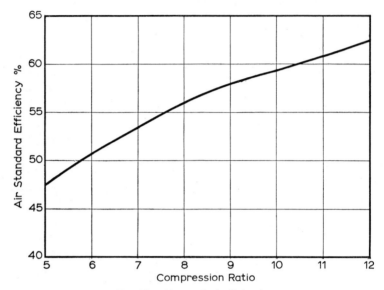

FIG. 56.   Air standard efficiencies

of the cylinder and the cylinder head and through the piston crown. A value of 1·3 to 1·35 is more representative of the coefficient for a typical petrol engine.

Yet another factor that prevents the practical engine from achieving the theoretical efficiency is the combustion phenomenon known as dissociation. At high temperatures and pressures a certain percentage of the oxygen molecules that have 'burned to completion' with the carbon molecules to form carbon dioxide (CO_2) tend to dissociate back to carbon monoxide (CO) and free oxygen (O_2).

$$2CO_2 \rightleftharpoons 2CO + O_2$$

Other reversible reactions occur; some of the hydrogen and oxygen molecules that have combined to form water vapour dissociate into hydrogen and oxygen again; certain nitrogen oxides also form and dissociate to a certain extent.

To the tuner the important things to remember are:

(*a*) dissociation means loss of useful work.

(*b*) the higher the compression ratio, the greater the amount of dissociation.

The combination of all the above factors drops the thermal efficiency of the petrol engine to about 65 per cent of the Air Standard Efficiency.

Mechanical efficiency

Power is produced inside the combustion chamber, but before it can be of any value the pressures exerted on the pistons must be transmitted to the crankshaft and flywheel. The efficiency of conversion of this energy to perform useful mechanical work at the flywheel is called *the Mechanical Efficiency*. All work performed in driving auxiliaries, such as the fan, the water pump, the oil pump, the generator and the distributor does not appear as useful work at the flywheel.

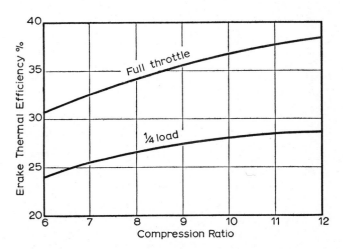

Fig. 57. Brake thermal efficiencies (*From 'The Private Car', 1960 Crompton-Lanchester Lectures, published by Institution of Mechanical Engineers*).

$$E_{\text{mech.}} = \frac{\text{b.m.e.p.}}{\text{i.m.e.p.}} \qquad (2)$$

where i.m.e.p. = the indicated mean effective pressure, or the mean pressure acting on the pistons during the working stroke.

b.m.e.p. = the brake mean effective pressure, or the equivalent mean piston pressure in terms of the load measured at the flywheel by a brake or dynamometer.

Fig. 57 gives values of the brake thermal efficiencies at different compression ratios for a typical high-speed petrol engine. By comparison with Fig. 56 we see how much of the fuel energy has disappeared in combustion losses and in mechanical losses. We also

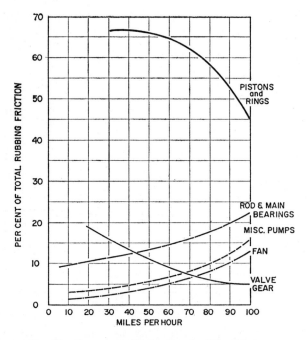

FIG. 58.   Percentage distribution of rubbing friction on Ford Falcon

see that when running at part-throttle conditions the mechanical losses represent a greater percentage of the total.

The losses that contribute to the reduction in mean effective pressure can be divided into two groups, mechanical friction losses and pumping losses.

Friction losses

In a typical modern o.h.v. engine of about $2\frac{1}{2}$ litres, the contribution of the various rotating and reciprocating components to the frictional total is given in Fig. 58. These values were measured on a Ford Falcon and were given in a 1961 S.A.E. paper by A. E. Cleveland and I. N. Bishop. They can be applied with reasonably accuracy to the case of a typical sports car engine. It is interesting to note how little power is lost, as a percentage of the total, in opening and closing the valves at 100 m.p.h. At all times the major friction loss occurs between the piston rings and the cylinder bores.

Pumping losses

Moving the gases in and out of the cylinders absorbs power. The losses may be considered as made up of two different physical effects.

(a) Work is required to compress the gas from the lower than atmospheric pressure in the induction manifold to the higher than atmospheric pressure in the exhaust manifold. At high induction vacuums the mass of gas to be compressed is low, but the pressure difference between the two manifolds is high. At low induction vacuums the mass of gas is greater, but the degree of compression from one manifold to the other is low. In general however, this particular pumping loss, expressed as a percentage of the total i.m.e.p. is greatest at part-throttle conditions than at full throttle.

(b) Work is also expended in passing the mixture through the various restrictions on the way to the cylinder and passing the exhaust gas through the various restrictions on the way out. These restrictions are:

1. the carburettor venturi,
2. the bends and constrictions in the manifold,
3. the bends in the head ports,
4. the inlet valve,
5. the combustion chamber walls adjacent to the inlet valve,

6. the combustion chamber walls adjacent to the exhaust valve,
7. the exhaust valve,
8. the exhaust ports in the head,
9. the exhaust manifold,
10. the silencer and tail pipe.

It will be seen that the pressure drop across the throttle plate is not listed under (*b*). It is the restriction produced by the throttle plate that determines the magnitude of the induction vacuum. The work expended in pumping the gases back to exhaust manifold pressure has already appeared under (*a*).

One item of work must appear on the credit side. As the hot exhaust gases leave the cylinder they contribute energy to assist in

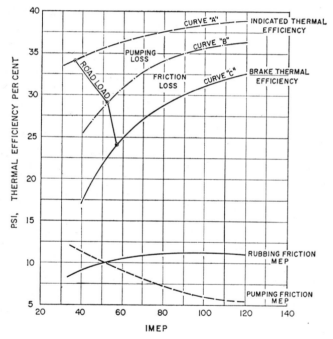

Fig. 59.   Friction and pumping losses on Ford Falcon at 1800 r.p.m.
Compression ratio is 10 to 1

their passage through the exhaust system. Moreover, with good design they can be made to contribute some of the energy required to induce the new charge into the cylinder.

It has been stated that pumping losses are greatest at small throttle openings, since the value of (*a*) rises when the induction vacuum is high. The values of Fig. 59 illustrate this.

Looking back over the preceding pages we can now see several ways in which more power can be extracted from an engine without burning one extra ounce of petrol. These are listed below.

TABLE 6

To Improve Brake Thermal Efficiency

(*a*) Increase the compression ratio.
(*b*) Modify the combustion chamber shape.
(*c*) Reduce gas flow losses by
    (i) fitting larger inlet valves;
    (ii) increasing port and manifold sizes;
    (iii) fitting larger carburettors, more carburettors or larger venturis in the existing carburettors;
    (iv) streamlining the throttle-plate. At full throttle it serves no useful function. A more streamline shape is therefore beneficial for this condition;
    (v) fitting larger exhaust valves;
    (vi) reducing exhaust system losses.
(*d*) Reduce friction losses.

Each of these methods will be discussed in detail later. In this chapter we are concerned only with items (*a*) and (*b*) from the above list.

The compression ratio

One can never make a precise prediction of the power gain to be made by an increase in compression ratio on a particular engine. For one thing, it is often difficult to increase the compression ratio without changing the basic shape of the head. A hemispherical combustion chamber is possible at 6 to 1 compression ratio, but at 10 or 12 to 1 C.R. the dome on the piston becomes so large and the depth of the chamber becomes so small that the shape of the chamber is closer to a piece of orange peel than a hemisphere. Taking another example, as the size of a typical push-rod o.h.v. combustion chamber is decreased to raise the compression ratio, the extent of the squish area, i.e. the area of the piston overlapped by the flat part of the cylinder head, increases. As this area increases, so does the squish turbulence effect at T.D.C. become more pronounced. Thus we see that other factors are inevitably involved in any change in compression ratio and prevent us from making any exact prediction of the outcome.

Experiments on the XO121 research engine, representative of a modern four-cylinder o.h.v. engine, reported by Gish, McCullough, Retzloff and Mueller in *S.A.E. Transactions*, Vol. 66, 1958, showed that the percentage improvement to be gained by a change in compression ratio is greater than predicted by the Air Standard Efficiency curve of Fig. 56. For example, the increase in power predicted from the Air Standard Efficiency formula if we increase the compression ratio from 5 to 1 to 10 to 1 is 26·9 per cent. On

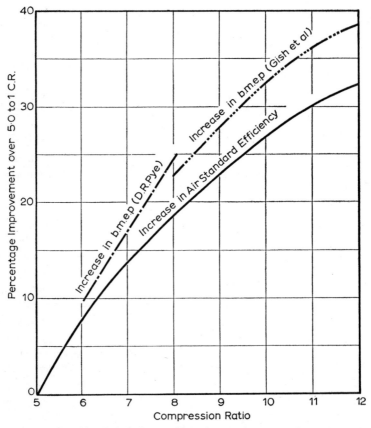

Fig. 60.   Gain in b.m.e.p. from increase in compression ratio

the XO121 engine a similar increase in compression ratio gave an increase in power of 32·3 per cent. A similar trend was found in an earlier set of experiments made by D. R. Pye on the Schneider Trophy Napier Lion engine. These results are given in Fig. 60.

With this graph we can make a reasonable estimate of the probable gain in peak power to be made by an intermediate change in compression ratio. For example, if the engine under consideration gives 150 b.h.p. at 8 to 1 compression ratio, the power at 10·5 to 1 compression ratio would be:

$$150 \times \frac{134 \cdot 3}{122 \cdot 9} = 163 \cdot 8 \text{ b.h.p.}$$

(The figures 134·3 and 122·9 are taken from the curve of Gish *et al.* in Fig. 60.)

Compression ratio limits

Limitations on compression ratio are set by two considerations: abnormal combustion and mechanical strength. Let us deal with abnormal combustion first. The reader will no doubt have a good working conception of the phenomenon we call knock, detonation or pinking. The petrol company advertisements have been trying to educate us for years on this subject. The use of higher and higher octane fuels has enabled us to increase the average compression ratio of sports car engines from 6·5 to 1 in 1940 to 9·0 to 1 in 1962. Experience with high compression engines, especially in America where ratios of 10·5 to 1 are now commonplace, has shown that the mere absence of true knock is no criterion of normal 'happy' combustion at these higher compression ratios. New words, such as 'thudding' and 'rumble' have been coined by the dynamometer engineers and are now entering the general jargon of the automobile engineer.

Abnormal combustion

The phenomenon of knocking combustion is well known to everyone in the motor trade and to most of the customers too. Even so, it is possible that some of us have got a rather muddled idea of the causes and the cure.

Knock

From the moment when the flame is first kindled at the sparking plug electrodes the spread of combustion into the main mass of the petrol-air mixture is initially relatively slow. Without the violent agitation, or turbulence, that exists in the mixture, combustion would be far too slow and would not be half-completed by the time the exhaust valve opened. Turbulence is present in the gas at all times, being generated and augmented as it passes through the carburettor venturi, through the manifold, around the inlet valve and finally increased to a new high as the piston crown traps a wedge of gas against the flat portion of the cylinder head and ejects this gas across the combustion chamber, the effect usually called 'squish'. While combustion is spreading like an expanding ball of flame with its centre at the sparking plug points, an exchange of heat energy is occurring between the burning charge and the unburned charge that lies ahead of the flame front. Heat is transferred largely by two paths, by radiation and by turbulent mixing of the gases. At the same time, some of the heat picked up by the unburned charge is given off again to the relatively cool walls of the combustion chamber to pass into the cooling water. All the time that this rapid exchange of heat is occurring there is a general rise of temperature and pressure in the combustion chamber. This rise is not uniform. The pressure is highest where the temperature is highest, i.e. immediately behind the flame front where combustion is more advanced towards completion. The temperature behind the flame front could be about 2000°C, while a short distance ahead of the front it could be as low as 500°C.

As combustion proceeds towards completion, however, the temperature and pressure of the unburned gas remaining in the cylinder becomes higher. Not only has this gas been receiving radiant heat for a longer period before the time of combustion but it has also been subjected to a wave of pressure that is now travelling ahead of the flame front. This final portion of mixture to be burned is usually called the 'end-gas' and it is in the manner of burning of this end-gas that the presence or absence of knocking depends. If by the time the flame front reaches this end-gas, the pressure and temperature of the end-gas has exceeded certain critical values for the particular air/fuel mixture, then the whole mass of end-gas will burn *instantaneously*, i.e. it will detonate or explode. If, however, the critical conditions are not reached burning will proceed progressively until the flame-front has spread across the entire mass of gas. When detonation occurs the pressure rise from the instantaneous combustion of the end-gas travels across the combustion chamber as a shock wave and produces a metallic 'tinkle' or 'ping' as it strikes the walls of the combustion chamber.

Knocking usually causes a slight loss of power, but this in itself is not a serious matter. Slight pinking during hard acceleration is quite acceptable on a high compression engine. More pronounced and prolonged easily audible knock cannot be tolerated. During knock more heat is transferred to the cylinder walls. Sparking plugs and exhaust valves operate at higher temperatures and more heat is rejected to the cooling system. Since knock is encouraged by high temperatures, the prolongation of knock only leads to more and more severe knock until finally failure of some overheated component such as an exhaust valve or a plug must inevitably occur.

The prevention of knock

Combustion of a fuel is a complex chemical process. Oxidation occurs in several stages and it is now known that the 'cooking' to which the end-gas is subjected during the combustion of the rest of the mixture results in the formation of certain pre-combustion products in the end-gas. Of these products, the ones that induce knocking combustion are the organic peroxides. Now we all know that certain fuels are more prone to knock than others. The fuels with the greatest resistance to knock are those with a molecular structure that is resistant to the formation of these peroxides, fuels such as methanol, iso-octane and other 'branched-chain' paraffins. A fuel such as normal hexane, or any other straight-chain paraffin, that forms peroxides relatively quickly when subjected to heat and pressure, is always prone to knock at moderate compression ratios. Benzene and toluene and their commercial mixture known as Benzole are ring compounds and exhibit a strong resistance to knock. Methyl and ethyl alcohol may be said to be knock-free, but under certain conditions they suffer pre-ignition (see below under pre-ignition). The addition of tetra-ethyl lead to petrol in minute quantities, as little as 2-3 cubic centimetres per gallon, produces a quite remarkable reduction in the undesirable pre-combustion

SCE N

reactions. It is used in all high octane pump petrols to boost the knock resistance of the base spirit, i.e. the 'unleaded fuel'.

The importance of the relative resistance to knock of different fuels was realised many years ago by both producers and users of fuels. A standard of comparison was evolved which is now universal. A sample of the test fuel is used to operate the C.F.R. (Co-operative Fuel Research) engine. This engine has a cylinder assembly which can be raised or lowered at will relative to the crankcase. This convenient method of varying the compression ratio is used to compare the performance of the test fuel with mixtures of a good reference fuel and a poor reference fuel. The good reference fuel is a particular iso-octane known to the fuel chemist as 2,2,4 tri-methyl pentane. The poor reference fuel is normal heptane. The mixture of these two reference fuels (or less expensive reference fuels of identical knock rating) that can be made to knock at the same compression ratio as the test fuel defines the 'octane number' of the test fuel. For example, if a mixture of 85 per cent iso-octane and 15 per cent normal heptane gives identical results to the test fuel, the test fuel would be rated as 85 octane. The above description of the technology of fuel knock-rating is sketchy in the extreme, but it will serve to help the tuner who is not acquainted with the subject to understand what is implied by the term 'octane number'.

Mechanical octanes

The American automotive engineers who invented this term were acknowledging the importance of cylinder head design in combating knock. 'Mechanical octanes' can be built into an engine by good design in such a way that 'fuel octanes' are saved. A good design of engine of 9 to 1 compression ratio will run knock-free on, say, 92 octane petrol, while a poor engine would need 98 octane fuel for the same compression ratio.

Other combustion abnormalities

Rumble. This is the name generally used to describe a low frequency noise sometimes heard on high compression engines at high speed and full throttle. Stating that some investigators have called it 'thudding', some 'pounding' and others simply 'roughness' should help the tuner to identify this noise when it occurs in engines in his care. Most investigators believe the noise to be caused by too rapid burning of the charge, the result of multiple ignition points, hot-spots in the combustion chamber or even by surface ignition from overheated carbon deposits. It is of interest to note however that the use of multiple *timed* ignition sources, such as the

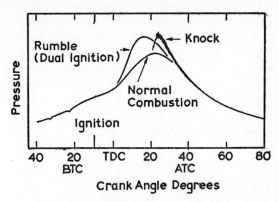

FIG. 61. Indicator diagrams of normal combustion, knock and rumble

use of two or more sparking plugs per cylinder, does not produce rumble. Indicator diagrams of normal combustion, knocking combustion and combustion with rumble are given in Fig. 61.

An increase in the octane rating of the fuel used does not usually cure rumble. The use of a blend containing more aromatics, such as benzol or toluol, often helps to reduce rumble. Some petrol companies now market fuels with phosphorus compounds as trace additives to reduce rumble and running-on. It is claimed that the additives fire-proof the carbon deposits and prevent surface ignition. In support of this it is interesting to note that rumble often occurs on modern engines when full throttle and high speeds are used immediately following a prolonged period of pottering at slow speeds in traffic. It is probable that the 'corn-flake' type of carbon deposits that builds up during several days of light load operation becomes overheated as soon as full load is used. Prolonged operation at full load would eventually break down these deposits, but there is a danger that pre-ignition will set in first.

Pre-ignition. This, as its name implies, is ignition of the mixture before the spark occurs. It can be caused by an overheated engine; it can be caused by incorrect timing; it can be caused by a hot-spot in the cylinder. Such a hot-spot could be a piece of glowing carbon, the edge of an exhaust valve, or, most common of all, an over-heated sparking plug. The actual ignition timing is thus un-controlled. Ignition occurs far too soon and the peak pressures occurring in the cylinder are sometimes excessive, giving a harsh metallic knocking sound, that can easily be mistaken for detonation. Since the cylinder pressures can sometimes be excessive before T.D.C. when pre-ignition occurs, the amount of 'negative work', i.e. the work done by the flywheel against the piston, is also ex-cessive and combustion is no longer as efficient as with a correctly timed spark. A drop in efficiency always means an increase in the amount of heat rejected to the cooling water and to the exhaust. Thus pre-ignition and overheating act in a vicious circle. The initial overheating may be caused because the combustion chambers are carrying an excessive amount of carbon, or an exhaust valve may be starting to burn, or sparking plugs of too low heat value may be overheating. Any of these causes will provide hot-spots that can lead to pre-ignition, which in turn by the uncontrolled over-ad-vance of the point of ignition will again raise temperatures still higher.

Pre-ignition is a dangerous condition and can lead to expensive engine failures. Typical failures are holes melted in piston crowns, seized pistons, burned exhaust valves.

Running-on. The failure of an engine to stop firing when the ignition has been switched off is more prevalent to-day than ever before. As a combustion phenomenon it is more related to rumble than to pre-ignition. Pre-ignition is ignition of the mixture *before* the sparking plug has fired. Rumble is caused by the presence of additional point sources of ignition that only reach a high enough temperature to ignite the mixture in their vicinity at about the same time as the firing of the plug. Running-on therefore occurs after the ignition has been switched off and one or more of these local hot-spots sustains combustion every time the fresh charge is compressed. The condition is not stable and the ignition timing can drift either earlier or later. A gradual drift to later ignition will cause incomplete burning, an occasional misfire and the eventual stopping of the engine. A drift to earlier ignition causes high and rapid rises in cylinder pressure, with rough running. The ignition eventually occurs so early that one of the pistons is stopped before T.D.C. and the engine starts to run backwards.

In all cases, rumble, pre-ignition and running-on, the cure is simple. Remove the hot-spot in the combustion chamber and the trouble will stop.

Combustion chamber shape

The combustion chamber is the heart of the engine. The world famous combustion expert, Westlake, even designs them in the shape of a heart! Despite the millions spent on combustion research to find the ideal shape there is still no conformity.

We can at least formulate a set of rules that can eliminate poor designs. The major factors that influence the knocking behaviour of an engine, the mechanical octanes that the designer can build into his engine, are as follows:

(a) the path to be traversed by the flame-front must be as short as possible.

(b) the distance between the sparking plug and the exhaust valve must be kept short.

(c) the end-gas must be in the coolest part of the chamber.

(d) the correct amount of turbulence must be designed into the head.

Types of combustion chamber

The shapes of combustion chamber used in modern sports car engines fall into three general types. Below we describe four types of chamber. The side-valve engine is not included. It is unsuitable for sports car engines and is in any case obsolete. The fourth type, although not used in any modern sports cars, is included for general interest, since there is no fundamental reason why it should not be used in a sports car engine. This type is the overhead inlet-side exhaust design as used on certain Rover, Bentley and Rolls-Royce engines.

I. *The 'bath-tub' head*

This is the most common shape in use to-day. It is used on single overhead camshaft engines and on push-rod o.h.v. engines. It has evolved from the simple flat cylindrical heads as used on heads with compression ratios of 5 and 6 to 1 in the early thirties. As com-

pression ratios have increased to the present-day values of 9 and 10
to 1 it has become necessary to confine the main clearance volume
into the form of an inverted bath-tub containing the two valve
heads at the bottom of the bath. The valve stems are sometimes set
at an angle of 10-15 degrees to the vertical. This serves a double
purpose. The angled valve port slightly improves volumetric

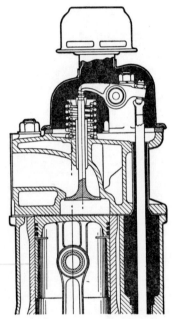

Fig. 62.   A typical bath-tub cylinder
head; the Triumph TR 3

efficiency and at the same time the slope of the bath-tub floor gives
a more suitable shape for smooth combustion. With such a shape
the progress of the flame front from the sparking plug as it expands
like a growing ball across the combustion chamber presents a
reduced volume of mixture for each tenth of an inch of radial
expansion. Thus a certain measure of compensation is provided
for the greater speed of the flame front in the middle part of the
combustion process, preventing too rapid a rise in cylinder pressure.
The final portion of the mixture is suitably cooled by the squish
area to prevent detonation.

II. *The wedge head*

The Coventry Climax FWA type head, as shown in Fig. 63, has a more pronounced slope to the chamber. This type of combustion chamber, which is very popular in America to-day, is best described as a wedge.

The chief disadvantage of the bath-tub head lies in the encroachment of the chamber walls on the valve heads. This gives a masking

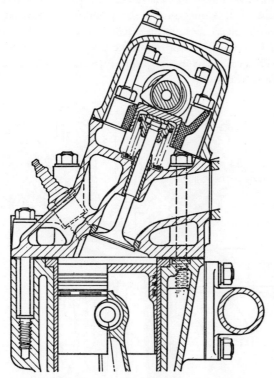

FIG. 63. Coventry Climax simple o.h.c. head

effect to the gas flow around approximately 40 per cent of the valve head. The masking effect is much reduced on the wedge type head and usually results in an improvement in breathing over the older design.

Both bath-tub and wedge heads are compact and can be designed to fulfil the four requirements for good combustion: (*a*) short

flame travel, (*b*) short distance from plug to exhaust valve, (*c*) a
well cooled end-gas, (*d*) the correct degree of turbulence.

It is to be noted that the squish area, the area of the head that
closely overlaps the piston crown at T.D.C. and by so doing causes
a jet of mixture to be injected into the main portion of the compressed gas shortly after combustion has begun, is generally made
smaller on high compression engines. The higher pressures and
temperatures produced in such heads result in higher flame speeds
and the use of the same amount of squish as on a low compression
engine would give excessive turbulence at T.D.C., the combustion
rate would be too high and excessive pressure rise would occur at
T.D.C. Physically this excessive pressure rise would be manifested
as rumble or roughness.

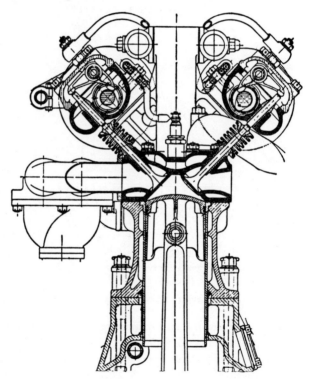

Fig. 64. The 1922 3-litre Vauxhall racing engine. A classic
design giving 129 b.h.p. from a compression ratio of only
5·8 to 1.

III. *The hemispherical head*

This is the classic shape of head as used in racing engines during the major part of the history of the petrol engine. Originally, with moderate compression ratios, the shape of the combustion chamber did indeed approximate to a hemisphere (see Fig. 64), but as compression ratios increased it became necessary to dome the piston crown more and more. To-day the modern 'hemispherical' head of 9 or 10 to 1 compression ratio is more like a new moon or a boomerang in cross-section (see Fig. 65). This is no

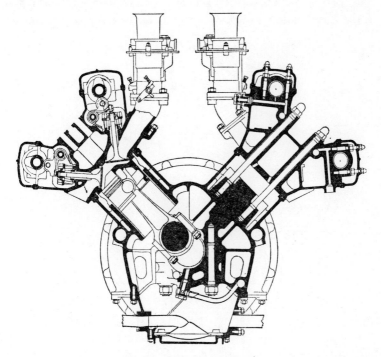

FIG. 65. Maserati, type 450 S

longer a compact design for efficient combustion, since the surface to volume ratio is far too high. At low engine speeds too much heat will be dissipated to the cooling jackets. Since the design is normally only used for high speed engines, this is no serious deficiency and is more than compensated by the excellent breathing afforded by this valve lay-out. The boundaries of the head do not encroach on the

valve heads, as in the bath-tub layout. The combination of the
angled bias of the valve ports and the unrestricted gas flow around
the full circumference of both inlet and exhaust valves gives the
hemispherical design undisputed position of Number One Power
Producer.

IV. *Overhead inlet-side exhaust head*

There have been several engines in the past using this design.
Coventry-Climax once made an 1100 c.c. engine of this type that
was used in the Morgan Four-Four. An excellent modern design
is the Rover shown in Fig. 66. The Rover design gives a compact

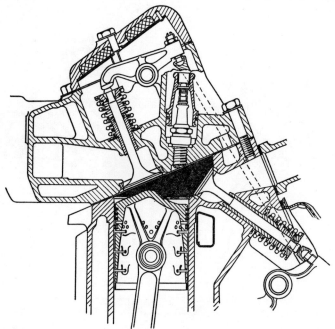

Fig. 66.   The Rover cylinder head

combustion chamber with the plug well placed. The inlet valve
head is not masked, as in the bath-tub layout, and the exhaust valve
is also well situated. Good cooling of the exhaust valve seat is
readily achieved. A very large inlet valve could be used with this
design. The actual limit in size is not dictated by the confines of
the head; only by valve inertia limitations.

From theory to practice

In this chapter we have confined our remarks to the simple theory of cylinder head design. The tuner is concerned with the application of this knowledge to his business of supertuning. The practical aspects of raising the compression ratio of an engine and the problems encountered will be described in Chapter Thirteen.

CHAPTER TEN

Volumetric Efficiency

THE POWER we can extract from a given size of engine depends largely on the amount of air we can make it breathe. 'Breathing' is a word we like to use in the world of tuning and we all know it is a useful way of describing the relative air consumption of engines, but the engineer prefers to be more explicit. Hence the ratio which we call 'the volumetric efficiency'.

Definition of volumetric efficiency

The volumetric efficiency is the ratio, expressed as a percentage, of the volume of air actually inspired by the engine per induction stroke (measured at standard temperature and pressure of 60°F and 14·7 lb. per sq. in.) and the nominal swept volume of one cylinder. No allowance is made for the apparent swept volume lost by the closing of the inlet valve after B.D.C. The true swept volume of piston area times stroke is the basis of comparison.

The volumetric efficiency is obviously low at part-throttle and will reach a maximum value at full-throttle at some point near the middle of the speed range. This speed will be fairly close to the speed of maximum torque. At higher speeds the volumetric efficiency falls, since the pressure drop between the atmosphere and the inside of the cylinder becomes quite considerable as the gas velocities through the inlet valves, induction system and carburettor venturi reach values in the range of 200-300 ft. per second. At maximum power the volumetric efficiency of a typical sports car engine would be about 75 per cent. At maximum torque engine speed it would increase to about 90 per cent. In certain cases the use of ramming induction pipes can increase the volumetric efficiency to nearly 100 per cent at the speed for which the pipes are tuned.

Better breathing

The more air we burn in an engine, the more fuel will be consumed too. If the mixture strength is maintained at the value that gives maximum power, the consumption of air and fuel will increase proportionately. Thus if we can get the engine to burn 20 per cent more air, it will also burn 20 per cent more fuel and will produce 20 per cent more power (perhaps slightly less than 20 per cent increase in power in general since the higher engine speeds involved carry a small penalty of higher friction and pumping losses). Of course, if the improved breathing is achieved by means of an

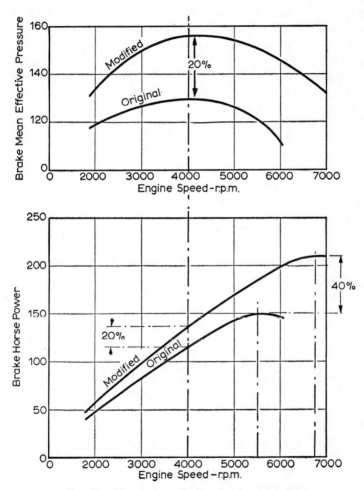

Fig. 67.   Max. power gain from improved breathing

engine-driven blower the power required to drive the blower must be deducted from the total.

Fig. 67 illustrates how improved breathing gives a greater increase to the peak of the b.h.p. curve than it does to the maximum b.m.e.p. If, for example, we fit larger carburettors, larger valves and a high lift camshaft and at the same time increase the compression ratio from 8 to 1 to 9·5 to 1, we might increase the maximum b.m.e.p. by 20 per cent (see Fig. 67 (*a*)). The b.h.p. *at the same r.p.m.* will also be stepped up by 20 per cent. The enhanced breathing, however, will shift the peak of the b.h.p. curve to a higher engine speed and the new peak b.h.p. will be about 40 per cent higher. This effect is shown in Fig. 67 (*b*).

As a rough guide we can expect that any improvement in breathing brought about by supertuning will increase the torque and power peaks according to the following law:

$$\frac{P_m}{P_o} = \left(\frac{T_m}{T_o}\right)^2 \tag{3}$$

where P_o = original maximum power
T_o = original maximum torque
P_m = maximum power after modification
T_m = maximum torque after modification.

The following table gives a list of recognised ways to improve the breathing of an engine.

TABLE 7

To Improve the Volumetric Efficiency

(*a*) Increase inlet valve size and lift.
(*b*) Modify inlet valve timing and opening period.
(*c*) Improve gas flow in cylinder head ports.
(*d*) Improve gas flow inside combustion chamber.
(*e*) Improve gas flow through induction manifold.
(*f*) Improve gas flow through carburettors.
(*g*) Increase exhaust valve size and lift.
(*h*) Modify exhaust valve timing and opening period.
(*i*) Improve exhaust gas flow in cylinder head ports.
(*j*) Improve exhaust gas flow in exhaust system.
(*k*) Supercharge.

A comparison of the two tables, Table 6 in Chapter Nine for Brake Thermal Efficiency and Table 7 in this chapter for Volumetric Efficiency, shows a certain amount of overlapping. For example, it is possible to carry out certain modifications to the carburettors, the induction system, the head ports and the inlet valves that bring about improvements, not only in breathing but in thermal efficiency. Sometimes, however, the requirements conflict and a measure of compromise is necessary. There is for instance a practical limit to the size of the carburettor choke. Too large a choke will give poor atomisation and combustion efficiency will suffer, giving a drop in brake thermal efficiency. Again it is possible to improve combustion efficiency by the use of inlet ports with a directional bias to give swirl turbulence during combustion. To induce this swirl there is a slight sacrifice in volumetric efficiency. An interesting example is given in the case history of the Jaguar XK engine. The XK 120 engine had biased inlet ports. The swirl turbulence helped this engine to give extremely good fuel consumption figures on test, yet the volumetric efficiency was still good, thanks to the use of a hemispherical head. In its later stages of development when the peak power had been stepped up by more than 50 per cent to 265 b.h.p., the biased port arrangement was dropped. The compression ratio had now risen to 9 to 1 from the original 7 to 1 and with the higher compression temperatures associated with the higher compression ratio the high degree of turbulence given by a biased port was no longer desirable.

THE INDUCTION PROCESS

On first thought it would seem that the induction process is largely a question of atmospheric air rushing into the cylinder to fill the void left by the descending piston. This however is an oversimplification. Let us consider the case of full-throttle operation at 3000 r.p.m. At the start of the induction stroke the cylinder is filled with the unexhausted combustion products, usually called 'the residual gases'. During the early part of the intake stroke the inlet valve is only open slightly and the rapid downward acceleration of the piston produces a near-adiabatic expansion of the residual gases with very little cylinder filling from mixture passing the inlet valve. As the inlet valve opening increases mixture rushes into the cylinder, but not at a high enough rate to match the piston displacement; so the cylinder pressure continues to fall. At some point before mid-stroke the rate at which the mixture flows past the inlet valve equals, then exceeds, the rate at which the piston is increasing the cylinder volume (this rate being equal to

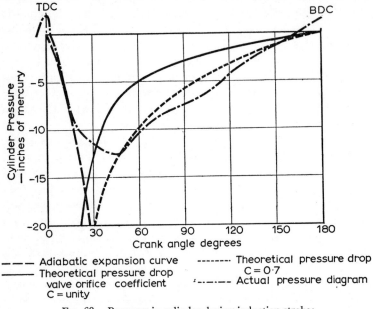

FIG. 68.   Pressures in cylinder during induction stroke;
full throttle at 3000 r.p.m.

$V_p \times A_p$, where V_p is the instantaneous piston velocity and A_p the piston area). From this point on, the pressure in the cylinder begins to rise and the real process of filling can be said to have begun.

As the piston approaches B.D.C. the piston displacement rate drops, but the mixture is still entering the cylinder at a high rate. It is therefore reasonable to delay the closing of the inlet valve to some point after B.D.C. until the inertia of the flowing gas stream in the induction tract is expended. The optimum point to close the inlet valve varies for different speeds and different throttle openings. The ideal for any particular condition would be to close it just as the gas flow into the cylinder ceases and just before the rising piston increases cylinder pressure to cause a back-flow into the induction tract.

Inlet Valve Timing

The choice of inlet valve opening and closing timings is, of necessity, a compromise. There is only one best timing for each particular running speed and load. An inlet valve timing that is best for 4000 r.p.m. will be too late for 2000 r.p.m. and too early for 6000 r.p.m. It is customary therefore to consider the type of service for which the engine is to be used when choosing the valve timing. This applies to the exhaust valve timing too.

The cam profile also exercises an influence on the choice of valve timing. With some very high lift cams, the opening and closing accelerations become so high that it becomes necessary to open earlier and close later than the optimum points in order to help reduce valve accelerations to acceptable values. Considered simply from the viewpoint of inducting the greatest possible charge, the inlet valve cam form would be one that opened the valve to full lift *instantaneously* at the optimum opening point and closed *instantaneously* from full lift at the optimum closing point. Such a cam form is obviously impractical since it would give rise to infinitely high loads on the valve train. We are therefore restricted in our openings and closings to the limiting accelerations that a valve train will stand without failure or excessive wear. In the conventional o.h.v. valve gear it is the valve spring (or springs) that maintain the tappet in contact with the cam face. During the closing phase of the operation the spring force at all times must never become less than the combined inertia forces of the valve, the rocker, pushrod and the tappet. One cannot, however, just fit an extremely stiff spring and solve the problem as easily as all that. During the opening operation of the valve the spring force is added to the inertia. The use of an excessively strong spring would therefore put such a load on the cam face that lubrication would break down and rapid scuffing would occur. To be free from cam scuffing contact pressures should not exceed 150,000 lb. per sq. in. Many modern camshafts, however, are asked to operate with pressures as high as 200,000 lb. per sq. in. Such designs are actually operating above the advised limits and one must not be surprised when wear does take place.

When valve lifts are taken to the limit, the good engine designer is forced to extend the opening period rather beyond his ideal opening and closing points. Modern camshafts are designed with

SCE O

long opening and closing ramps in order to help reduce opening and closing accelerations. Thus it is possible for the actual valve closing time to be as late as 75° after B.D.C., but for the last 20° of closing movement the valve is only a few thousands of an inch away from its seat and the amount of gas that can back-flow is negligible.

A typical touring type engine such as the Austin-Healey Sprite has a valve timing of I.V.O. 5° B.T.D.C.; I.V.C. 45° A.B.D.C.; E.V.O. 40° B.B.D.C.; E.V.C. 10° A.T.D.C. Both inlet and exhaust valves have therefore opening periods of 230°. By sacrificing some low speed torque the same basic cam profile, giving identical valve

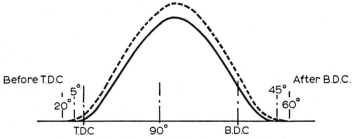

FIG. 69.   Greater valve lift for the same accelerations

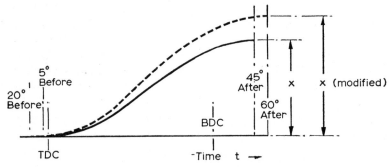

FIG. 70.   The valve area integral

accelerations, can be used to give a greater valve lift and a much greater total valve area. How this is done is shown by the two diagrams of Fig. 69 and 70. The first represents the two cam profiles in terms of lift per degree of crank rotation. The example chosen is the inlet cam, but the same technique would apply also to the exhaust cam. The modified cam is shown as opening 15 degrees earlier and closing 15 degrees later than the standard cam. This extends the opening period by about 13 per cent. With this modified valve timing the lift can be increased by 13 per cent without any increase in valve accelerations or cam face loads at corresponding engine speeds. Thus the Austin-Healey Sprite valve lift can be increased from 0·280″ to 0·316″ with this modified timing without incurring any increase in valve train loadings *at the same engine speed*. A note of warning is necessary here. An increase in valve area integral, i.e. the summation of area times 'time', of about 28 per cent will be given by the new cam and the much improved breathing will move the peak of the power curve to a much higher speed. This will inevitably increase the stresses in the valve train and the pressures on the cam face. If we fix an r.p.m. limit that allows for a maximum engine speed 15 per cent higher than standard, the valve train stresses will increase by 32 per cent.

The Iskenderian MM-55 Track cam for the Sprite has a valve timing of 20; 60; 60; 20 and a lift of 0·325″. The lift is increased by 16 per cent from standard and the valve opening period by 13 per cent. As a rough guide we can add the two figures 16 and 13 and say that the valve opening integral will be increased by about 29 per cent, giving 29 per cent improvement in breathing. This assumes of course that the cam profiles conform to the same acceleration formulae. It is interesting to note then that the fitting of this Isky cam steps up the peak of the power curve from 5200 r.p.m. to 6750 r.p.m., an increase of 30 per cent.

Stronger valve springs are required to resist the increased inertia loads at these high rotational speeds. Increased cam face wear is almost inevitable with such a camshaft, but Ed. Iskenderian minimises this by the use of the latest techniques of surface hardening.

Overlap

All modern engines have a valve timing in which the inlet valve opens before the exhaust valve closes. The extent of this period when the induction tract and the exhaust manifold (in theory, at least) are in direct communication by way of the combustion chamber is usually called 'valve overlap'. With a small overlap, i.e. 20° or less, the communication between the two ports is almost

non-existent. At the middle of the overlap period a typical engine
with a 20° overlap would have no more than 0·012″ of opening on
each valve. At speeds above 2000 r.p.m. at full throttle the back-
flow of exhaust gas into the induction tract must be negligible with
such small openings. At idle and with a nearly closed throttle at
medium speeds the tendency for a back-flow of exhaust gas is
increased, since the depression in the induction tract is high at such
times and will thus encourage the flow of exhaust gas from the
region of higher pressure inside the cylinder. With a larger overlap,
i.e. 40° or more, both inlet and exhaust valves will be off their
seats by 0·030″ or more at T.D.C. and a much more effective
connection exists at this time between the exhaust and induction
systems. This is a very important factor when a 'tuned' exhaust
system is used. The use of a 'tuned length' of exhaust pipe to assist
in the extraction of the exhaust gas from the combustion chamber
and to induce a larger flow of fresh charge into the cylinder will be
discussed at length in Chapter Eleven. It is important at this stage
to remember that the use of a large overlap will permit full benefit
to be taken of this extraction effect, *but only over a limited range of
engine speeds*. At certain other speeds there will be a tendency for a
back-flow of exhaust gas into the combustion chamber and across
the combustion chamber into the induction system. Thus a gain
in power at one engine speed is balanced by a loss in power at
another. A secondary disadvantage is sometimes apparent when
an unsymmetrical induction system is used, since the exhaust
back-flow will sometimes be greater for some cylinders than for
others. In such cases mixture distribution can be seriously affected
at some engine speeds. In general then, large overlaps of 60° or 80°
should be confined to engines with one carburettor choke per
cylinder. Such engines seldom give a smooth idle below 700 r.p.m.
since excessive amounts of exhaust gas flow back into the induction
ports when the induction vacuum is high. This contaminates the
fresh charge to such a degree that the oxygen content approaches
the lower limit for the support of combustion.

INLET VALVE AND THROAT DESIGN

Use of the air flow rig

The design of the inlet valve, its throat and the port in the
cylinder head is still to some extent an empirical art. A scientific

approach is seen in the modern use of flow rigs but, even here, one starts with the assumption that the valve/throat/port arrangement that gives the greatest flow under steady flow conditions will also give the greatest flow under the pulsating flow conditions of engine operation. Occasionally this assumption can lead us into error.

An air-flowed head may be a fine status symbol when competing at the club bar, but to the automobile engineer it is simply a technical short-cut—an attempt to simulate the real thing—the measurement of power on the dynamometer. Not all the factors involved can be simulated. Air-flowing is cheap—but it cannot replace engine development on the dynamometer.

The earliest air-flowing in the writer's experience was more direct, in that the measurements were actually made on the test-bed with an engine under power. Such work can be very laborious. Every change, whether it be in valve design or port shape or head shape, calls for hours spent in dismantling and reassembling after every change. One can learn so much from the results. With readings of b.m.e.p., fuel consumption and air consumption available one can soon decide whether a modification has improved combustion efficiency, volumetric efficiency or both.

The modern technique of air-flowing involves the making of wooden models of the cylinder head, complete with valves, but neglecting unrelated details, such as water jackets. Sometimes an actual cast-iron head will be used, but wood will be used to build up different shaped valve throats and ports. With these models we can measure the relationship between pressure drop and air rate when air is passed through the system. This can be done in several ways. A source of compressed air at low pressures (sometimes as low as 8 inches of water) can be blown through the inlet port with the valve in the fully opened position. Then with a fixed pressure upstream of the valve the air rate can be measured, or, alternatively, with a fixed air rate the drop in pressure across the system can be measured. Another alternative is to use a vacuum pump on the downstream side as shown in Fig. 71.

With any of the above methods much more can be learned if the air-flow readings are taken for a range of valve openings in steps of 10 degrees of crank rotation. The valve lift data must be known to use this method accurately, the corresponding valve opening for each step of 10 degrees being carefully measured by a micrometer

gauge as shown in Fig. 71. This step-by-step air-flow data is most
essential when the value of venturi ports is being considered. A
change to venturi ports can sometimes improve the flow up to some
point short of full valve opening and decrease it after this point.

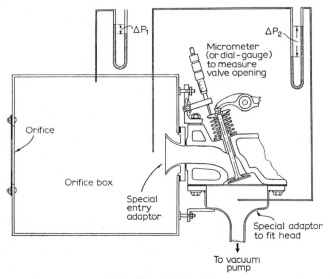

FIG. 71.   Air flow rig

Once the flow rig is made and the technique mastered, port and
valve throat changes can be made quite rapidly by carving out
pieces of wood or building up different shapes with plastic wood
and re-checking the flow rates until the tuner is finally satisfied that
he has achieved the best design.

A simple design of air flow rig

The rig shown in Fig. 71 was used in the engine labs of the
Massachussets Institute of Technology for many experiments on
valves, valve seats and throats. It will serve admirably as a basis for
a description of the method.

If the pressure drop across the system is kept small we can, without
introducing any great error, consider the air as a incompressible
fluid. For this condition we can use the simple flow formula:

$$M = A \, C\sqrt{2\rho \, \Delta P \, g} \qquad\qquad (4)$$

where M = mass flow, lb./sec.

      A = valve area, ft.²

      C = orifice coefficient

      ρ = mass density of air, lb. per ft.³

    ΔP = pressure drop, lb. per ft.²

      g = gravitational acceleration, ft. per sec.²

In applying the above formula one must decide on which area one will base the coefficient C. This could be the area of the valve seat, of the throat, or of the valve head. The valve head area is a

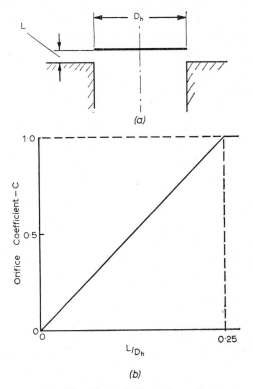

FIG. 72.   Theoretically perfect valve

logical basis, since the size of valve head that can be accommodated in a given cylinder head is a very practical design limitation. Using this basis, as shown in Fig. 72 (a), where D_h represents the valve head diameter and L is the valve lift, we can now predict what value of C would be given by a theoretically perfect valve, i.e. a valve with an orifice coefficient of unity at full lift.

The area of the discharge opening is given by:

$$A = \pi L D_h$$

This is equal to the area of the valve head when $L/D_h = 0.25$.

Thus for greater lifts than $0.25\ D_h$ there is no increase in effective valve area. The curve of Fig. 72 (b) thus provides us with a theoretical perfect standard with which to compare actual valve and port data.

Flow calculations

Referring again to Fig. 71, we see that the mass flow of air through the valve/port combination to be tested also passes through the metering orifice. By equating these in terms of expression (4) we obtain the following:

$$M = A_o\,C_o\sqrt{2\,\rho_1\,\Delta P_1} = A_v\,C_v\sqrt{2\,\rho_2\,\Delta P_2} \qquad (5)$$

where
M = mass flow, lb./sec.
A_o = area of measuring orifice, ft.²
A_v = area of valve head, ft.²
C_o = discharge coefficient for orifice (approximately 0.59 for sharp-edged orifice)
C_v = discharge coefficient for valve
ρ_1 = mass density of air before orifice, lb./ft.³
ρ_2 = mass density of air before valve, lb./ft.³
ΔP_1 = pressure drop across orifice, lb./ft.²
ΔP_2 = pressure drop across valve, lb./ft.²

Since ρ_1 is very nearly equal to ρ_2

$$C_v = \frac{A_o\,C_o}{A_v}\ \sqrt{\frac{\Delta P_1}{\Delta P_2}} \qquad (6)$$

This is a very simple formula to apply, since manometer readings only need be taken at each setting of the valve opening.

The flow rig has been a wonderful tool in the hands of men like Weslake and Goulding and for the man who cannot afford the expense of a dynamometer it can give good guidance to the professional tuner who wishes to develop a tuning kit or a special cylinder head for sale to the public. But what of the man who runs a tune-shop where almost every job is different? Flow testing is, in general, far too expensive for one-off jobs. The small-time tuner just has to try to be smarter than the big people. It's not easy, but it has been done. The small man must rely on experience and what he lacks in technical knowledge he must pick up from books.

The writer has been on both sides of the fence and is well aware of the problems of the small-time garage tuner. For the help of all such people then, all those who have neither test-bed nor flow-bench, let us try to set down on paper a few simple design rules, draw a few graphs to assist these one-off tuners in their choice of valves and port sizes. Even the engine designer must have certain signposts to guide him when he lays out his preliminary sketches.

In discussing valves and valve throats one often encounters the terms 'mean gas velocity at the throat' and 'mean gas velocity at the valve'. The gas velocities during the induction stroke vary from zero to a maximum and back again to zero. To measure these changes in velocity accurately at any part of the port or valve profile would be a major piece of research on any engine and with this in mind it has been the custom in design circles to make the following simplifications to provide an easily calculated basis of comparison. The gas velocity through the throat (indicated by dimension D_t in Fig. 73) and between the valve and its seat in the head (shown at a in Fig. 73) is related by simple mathematics to the volume swept by the piston during one stroke. Thus no allowance is made for the influence of valve timing, valve overlap or exhaust-induced mixture, even for the temperature of the gas.

The mean piston speed at N r.p.m.,

$$V_p \text{ (ft./sec.)} = \frac{2S}{12} \times \frac{N}{60} \tag{7}$$

where S = the stroke in inches

The velocity through the valve throat, of diameter D_t inches, can be directly related to the piston speed by the ratio piston area/throat area, $\left(\dfrac{D_p}{D_t}\right)^2$, where D_p is the piston diameter. (It is conventional to make no allowance for the reduction in throat area caused by the valve stem.)

The mean gas velocity at the valve throat,

$$V_d \text{ (ft./sec.)} = \frac{SN}{360} \times \left(\frac{D_p}{D_t}\right)^2 \qquad (8)$$

Similarly, one can equate the piston velocity to the area exposed between the valve and its seat at full valve lift. But why only full lift? Again we can only say this is a convention and it has served remarkably well in the past for making quick comparisons between different designs of engine. The valve area in question is really a

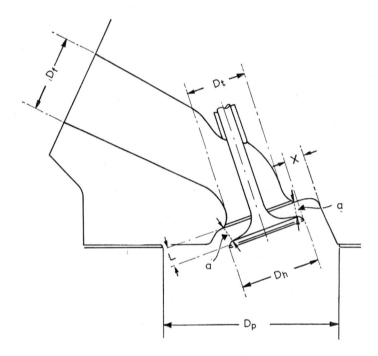

FIG. 73.   Valve and throat dimensions

truncated cone taken at the point *a* in Fig. 73 where the minimum gap occurs between valve and seat at full lift. No great error is involved if we simplify this area to the circumference of the throat, πD_t, multiplied by the maximum lift, *L*.

The mean gas velocity at the valve,

$$V_v \text{ (ft./sec.)} = \frac{SN}{360} \times \frac{\text{piston area}}{\text{valve opening area}}$$

$$= \frac{SN}{360} \times \frac{\frac{\pi}{4} D_p^2}{\pi D_t L}$$

$$= \frac{SN}{360} \times \frac{D_p^2}{4 D_t l} \qquad (9)$$

In Table 8 below we have calculated values of V_t and V_v for several well-known sports car engines fitted with bath-tub or wedge combustion chambers. Values for three hemispherical head engines are given in the section (b) of the table.

TABLE 8
INLET VALVE AND THROAT GAS VELOCITIES

Engine	Valve head dia. inches D_h	Throat dia. inches D_t	Valve lift inches L	$\frac{L}{D_t}$	Mean gas velocity at max. power r.p.m. ft./sec.	
					at throat V_t	at valve V_v
(a) Bath-tub and wedge type heads						
A.-Healey Sprite	1·094	0·968	0·28	0·289	284	246
A.-Healey 100-6	1·688	1·378	0·357	0·258	250	242
B.M.W. 507	1·50	1·32	0·315	0·238	259	271
MGA 1500	1·50	1·25	0·357	0·286	296	259
(b) Hemispherical heads						
Coventry Climax FPF	1·75	1·45	0·35	0·242	275	284
Jaguar XK 120	1·75	1·55	0·312	0·202	310	329
Jaguar, D Type	1·875	1·60	0·375	0·235	284	302

The superior breathing of the hemispherical head is shown up clearly by the higher gas velocities at the inlet valve. In the case of the wedge head, and to a greater extent for bath-tub head, the cylinder head tends to mask a section of the valve head. Even when the distance x (see Fig. 73) is adequate the gas stream is deflected through a considerable angle and the resistance to flow is greater than with the hemispherical head.

Based on a mean value of $V_t = 260$ ft./sec., taken from Table 8 (*a*), the graphs in Fig. 74 should serve as a guide to the tuner when he is calculating the breathing potential of an engine. Let us take an example: we have carried out work on a 2 litre 4-cylinder engine involving the fitting of much larger inlet valves. The valve throats have been increased from 1·6 inches to 1·8 inches in diameter. The compression ratio has been increased from 8 to 1 to 9 to 1 and the carburettors have been changed from two single-choke to two twin-choke of the same choke size. After all this work we find that the power has not increased by more than 10 per cent and the peak of the power curve seems to have moved from 5800 r.p.m. to 6200 r.p.m.; a disappointing increase in performance. For a cubic capacity of 500 c.c. per cylinder we see from Fig. 74 that an inlet

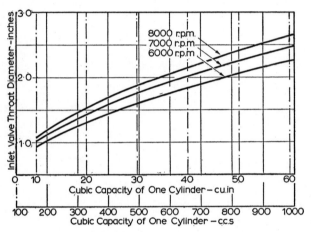

FIG. 74.   Inlet valve throat sizes
(Bath-tub and wedge type heads)

throat size of 1·8 inches should give a peak power of approximately 7200 r.p.m. We now know that some form of restriction exists that is preventing the engine from realising the true potential flow of the new larger valve throats. It could exist in the carburettors, in the ports, the induction pipes, the exhaust valves, exhaust valve throats, manifolds, silencers, tail pipes. The valve lift could be inadequate, the valve timing unsuitable; even the combustion chamber could be restricting the gas flow round the valves.

This latter failing is not uncommon. The fitting of larger valves in a bath-tub head reduces the dimension x in Fig. 73. The combustion chamber wall can usually be machined to an increased radius, usually about $\frac{3}{32}$ inch greater, but there is no point in increasing valve diameter at the expense of a serious restriction here. A useful guide to a minimum dimension for x is to maintain this at no less than 0·7 times L, the valve lift.

With a well designed hemispherical head the walls of the combustion chamber should not mask either the inlet or the exhaust valve. From Table 8 (b) it will be seen that higher inlet valve velocities are achieved at peak r.p.m. with the hemispherical engines. This is because the distribution of gas velocities around the periphery of the valve head is usually very regular on these engines. On the bath-tub head it will vary from 200 ft./sec. approximately where the combustion chamber wall encroaches to 300 ft./sec. where no restriction occurs. The gas velocities on the now obsolete side-valve engine seldom exceeded 200 ft./sec.

Carburettor venturi size

The carburettor size must be well matched to the inlet valve size if we are to get the best performance from an engine. If the carburettors are too small for the valves the desired peak r.p.m. will not be reached. If the carburettors are too large, torque at low speed will suffer.

The author has found the following formula useful in choosing venturi sizes. Naturally, there is no exact formula that will give the correct size for all engines, from four cylinders to sixteen cylinders, from long strokes to short strokes, but the following formula will serve as a starting point. It will be noted that the number of cylinders and the number of carburettors do not enter into the expression.

$$\text{Venturi size, m.m.} = 20 \sqrt{\frac{V}{1000} \times \frac{N}{1000}} \tag{10}$$

where
$\qquad V$ = the swept volume of one cylinder, c.c.
$\qquad N$ = the required r.p.m. at peak of power curve.

This is based on a mean gas velocity through the venturi of 360 ft./sec. at the peak of the power curve.

Thus if the valve throat size is already determined we should

make the venturi size $\sqrt{\dfrac{260}{360}}$ of this size, or 85 per cent, in the case

of a bath-tub head and $\sqrt{\dfrac{300}{360}}$ of the throat size, or 90 per cent, in

the case of a hemispherical head.

Valve seats

Designers still do not agree on the question of seat angles. The old M.G. XPAG series engine used a seat angle of 30° (120° included angle) for both inlet and exhaust valves. The MGA engine uses a 45° seat angle for both inlet and exhaust valves (from engine no. 4045 onwards an interference angle was given to increase the speed of valve bedding. Thus the seat in the head was 45° and the seat on the valve was $45\tfrac{1}{2}°$.) The Triumph engine designers also prefer seat angles of 45° for both valves, but the Austin-Healey designers are of two minds, using 45° for both inlet and exhaust on the Sprite engine and 30° for inlet and 45° for exhaust on the larger '3000' engine.

The flow characteristics of both inlet and exhaust valves have been investigated by several research workers. All agree that a 45° exhaust valve seat gives the lowest pressure loss. There is no measure of agreement on the question of the inlet valve seat angle. The writer ventures to give the following as a guide to the most efficient angle. Where the valve lift is relatively low, i.e. the ratio of lift to throat diameter is less than 0·25, use a 30° seat. Where the lift is high, use a 45° seat.

The width of the valve seat is an important design feature and again it is difficult to generalise. Inlet valve seats are usually made about 0·040″ wide per inch of valve head diameter. Exhaust valve seats must be wider in order to provide the area for heat dissipation. An average design value would be 0·060″ per inch of valve head diameter, but with an upper limit of 0·100″. The use of too wide a seat would be just as disastrous to the exhaust valve as too narrow a seat. Lead deposits from the T.E.L. in the fuel tend to adhere to the exhaust valve and a seat that was too wide would fail to bite through these deposits when the valve returned to its seat. Differential seat angles have given excellent service on supercharged acro engines and are well worth the trouble on high performance auto-mobile engines. The difference between the angle on the valve seat and the angle on the seat in the head can be anything from $\frac{1}{2}°$ to 2°, the smaller angle being in the head so that the initial contact between the two seats is on the outer seat diameter.

Inlet valve throat and port design

Some confusion of terminology exists in the use of the terms 'throat' and 'port'. Some automobile engineers use the term 'port' to include the passage in the cylinder head all the way to the valve seat. Some call the port only that portion of the passage adjacent to the valve seat. Others call this the throat. For the sake of clarity, within the confines of this book at least, let us define the valve port as the passage in the cylinder head extending between the manifold face and the particular portion of the gas passage that terminates where the passage begins to expand to the valve seat. This 'port end' we will call the 'throat'.

The valve throat

The influence of the ratio of valve throat diameter to valve head diameter was investigated by R. G. Falls and S. W. James (M.I.T. Thesis, 1941: 'Effect of Diameter Ratios on Flow through Inlet Valves of Internal Combustion Engines'). Curves for different ratios are shown in Fig. 75. In every case the radius of the curve

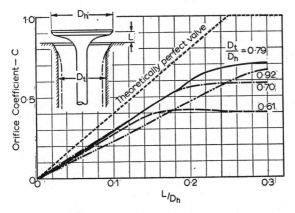

FIG. 75. The influence of the valve throat

joining the parallel portion above the throat, i.e. the end of the port, and the seat in the head was made so that the curve joined the seat at a tangent. The closest approach to the theoretically perfect line (shown by dotted line) is given by

$$\frac{D_t}{D_h} = 0.79 \qquad (11)$$

Throat shape

The influence of inlet throat shape was investigated by D. U. Hunter as long ago as 1938 (M.I.T. Thesis 1938: 'The Effect of Valve and Port Shapes on Air Flow through Intake Valves'). Hunter used the modern air-flow technique of building up port

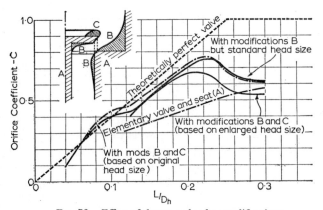

Fig. 76.   Effect of throat and valve modifications

and throat shapes in Plasticine, adding fairings to the throat and to the valve head, both upstream and downstream of the direction of flow. The results are shown in Fig. 76.

The fairing C appears to make an improvement when the orifice coefficient is compared with the original head diameter. This is not a fair comparison, however, since the use of such a fairing increases the effective valve head diameter. This, it must be remembered, is a design parameter we chose from the start, since the size of valve head is usually limited by the available space in the combustion chamber. The corrected curve, based on the enlarged diameter, shows an increase in coefficient up to an L/D_h ratio of 0.15 and a decrease at higher ratios. Since the provision of such a fairing would increase the valve weight considerably, it is not recommended.

Valve radius under head

It is generally believed by the tuning fraternity that a big radius under the inlet valve will give an improved gas flow and the only reason we have to compromise here is to keep the weight of the valve down. Hunter's experiments nevertheless showed that a really large radius reduces the flow (see Fig. 77). The large radius

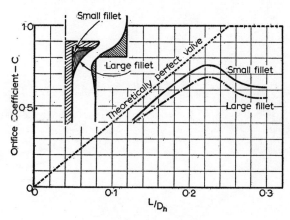

FIG. 77.   The influence of valve head fillet

is equal to $1\frac{1}{4}$ times the valve head radius; the small radius is equal to $\frac{1}{4}$ times. It should be noted that the small radiused 'fairing' is supplemented by a 10° slope to the back of the valve.

Port shape

Since our aim is to get the charge of mixture into the cylinder with as little pressure loss as possible it is hardly necessary to elaborate on such obvious bad features as sharp bends and needless changes, either up or down, in cross-sectional area. Most designers of high performance engines attempt to achieve an induction pipe that is as near to straight as possible and one that does not vary in cross-sectional area, all the way from the atmosphere to the cylinder bore. On hemispherical head engines, with very big valves, one can approach this ideal very closely (see Fig. 78). On bath-tub

SCE P

and wedge type heads the valve size is smaller than the maximum carburettor size we can usually use with it. In such cases it is a good trick to use this large bore carburettor and provide a gradual decrease in cross-section all the way from the carburettor to the valve throat (neglecting of course the particular restriction of the carburettor venturi). If the included angle of the sloping walls does not exceed 8 degrees it is possible to give a gradual acceleration to

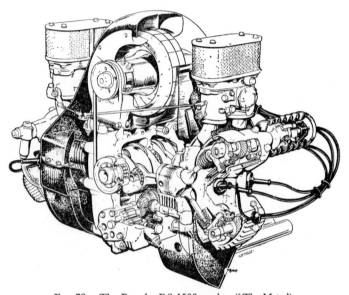

FIG. 78.   The Porsche RS 1500 engine ('*The Motor*')

the gas stream, without inducing any additional turbulent eddy losses from the gradual change in cross-section. For example, an included angle of 5 degrees will reduce the port diameter from 2 inches at the carburettor flange mounting face (XX′ in Fig. 79) to $1\frac{1}{2}$ inches diameter at the throat, over a port length of about 6 inches. With a throat velocity of 260 ft./sec. at maximum power r.p.m. the gas velocity at XX′ would only be 146 ft./sec. The use of air intake trumpets of 3 inches diameter would drop the entry velocity to 65 ft./sec. at maximum power r.p.m.

In this way we can achieve a gradual acceleration of the column of gas and take us one step further in our attempt to get the gas into the cylinder at minimum loss of pressure. As explained earlier, the figure of 260 ft./sec. for the throat gas velocity at maximum power r.p.m. is a nominal mean gas speed over the whole induction stroke. The gas velocity through the partially closed valve at B.D.C. will be about 500 ft./sec. By converting the kinetic energy of this high velocity column of gas into static energy we can gain the equivalent of 2 lb. per sq. in. in increased cylinder pressure. This

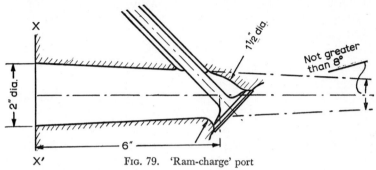

FIG. 79. 'Ram-charge' port

is the effect recently christened 'ram-charge' by the Chrysler Corporation and used to such good effect on Dodge cars. All engines exhibit this phenomenon to some extent, but the use of a gradually diminishing cross-section in the induction system plus an induction system of tuned length (to be discussed more fully in Chapter Eleven) can throw a satisfying hump into the power curve just where we usually want it—right in the middle of the speed range.

Venturi ports

Some designers make a small constriction at the point where the throat runs into the port. There is little doubt that this helps to raise the level of turbulence at low engine speeds, but the designer or tuner who is trying to get the utmost in r.p.m. from an engine usually avoids the venturi port. The Jaguar XK 120 engine had a small venturi effect to the inlet ports, but the port was also offset relative to the cylinder axis to give a directional swirl to the mixture.

The use of a slight venturi behind the throat is almost inevitable when valve inserts are fitted, as in alloy heads. Figs. 63 and 66 in Chapter Nine show how a reduction in diameter towards the inner face of a valve insert serves to give the required wall thickness to the insert at this point. Without this waisting effect a larger outside diameter would be needed for the insert for the same valve head diameter. A 30 degree seat angle is usually used with a venturi port since the gas flow in the region of the seat has a pronounced outward component.

Curved ports

The ports must always have some curvature, since they must be deflected below the valve operating gear. In some hemispherical head engines such as shown in Figs. 65 and 78, the designer has been clever enough to reduce this deflection so much that the port is almost straight. How much has the designer gained in this way, we may ask, and a partial answer can be found in the experiments carried out for the Ranger Engineering Corporation at the Massachussets Institute of Technology by George L. Estes, Jr., and James E. Hawkes. These experiments were directed to the practical problem of developing the breathing potential of an aero engine

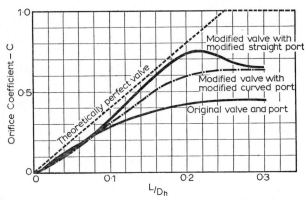

FIG. 80. Air-flowing the Ranger cylinder head

cylinder head. It was significant that an early comparison between a straight port and a curved port, when the design was still as it had left the drawing board and the system was still a relatively inefficient breathing device, showed a curved port, with a fairly sharp deflection of 90°, to be just as good as a straight port. When the system was in an advanced state of development, the provision of a bend in the port clearly showed a greater pressure loss. Fig. 80 shows the actual orifice coefficients measured over the range of valve openings. At an L/D_h ratio of about 0·28 the straight port was hardly any more efficient than the elbowed port, but at smaller valve openings, the straight port was much more efficient and the overall effect would be an improvement in air flow of about 10 per cent. The moral then is simple. Make the port as straight as humanly possible within the mechanical limitations imposed by the valve operating gear.

THE EXHAUST PROCESS

In theory the maximum combustion efficiency is given when the exhaust gases are allowed to expand to the full extent of the exhaust stroke. In practice, even on a touring type engine, the exhaust valve opens no later than 40° before B.D.C. To delay the opening of the exhaust valve any later would involve a greater amount of negative work to be performed by the piston on the exhaust gases during the exhaust stroke than the amount of positive work gained on the down stroke by the extension of the valve opening point. If the valve were to open at B.D.C. the piston would be about one-third of the way back towards T.D.C. before any appreciable exhaust valve opening had occurred. This would create a considerable back-pressure inside the cylinder, making the piston perform work on the crankshaft and flywheel in the reverse direction, i.e. negative work. Here, of course, we are considering an extreme case. Even at very low speeds such a timing would be inefficient. In general, however, the choice of exhaust valve opening time, is related to the particular operating speed of the engine. A low speed engine with a power curve peaking at about 4500 r.p.m. would have an exhaust valve opening at about 40° before B.D.C. and very little overlap at T.D.C., i.e. an exhaust valve closing of about 10° after T.D.C. For a high speed competition engine, however, with a power curve rising to 7000 or even 7500 r.p.m., power must be sacrificed at the bottom of the expansion curve in order to get the exhaust valve opened early in the exhaust stroke. For such an engine an exhaust valve opening of 70° or even 75° before B.D.C. would be appropriate. With a large overlap, as one would most certainly use if the engine were provided with one choke per cylinder, the exhaust valve closing would be 30°, 35° or even 40° after T.D.C. Compared with the slow-speed engine timing of 40° before B.D.C. to 10° after T.D.C. the extension of the timing at both ends to 70° before B.D.C. and 40° after T.D.C. would increase the duration of exhaust valve opening from 230° to 290°, an increase of 26 per cent.

Unfortunately the use of large opening durations on an exhaust valve carries a certain disadvantage—a very important one too. The longer the valve stays off the seat, the less time it spends in contact with it. This is self-evident. While the valve is off its seat the valve head is picking up heat; while it is in contact with its seat it is giving up heat to the cylinder head and the cooling water. Thus a reduction in contact time, as a percentage of the whole, means a hotter running exhaust valve. In some engines this factor can place a design limit on the valve opening duration. The exhaust valve has always been a serious limitation to power output. Indeed when we are considering the further supertuning of an already highly modified engine the exhaust valve becomes one of the most serious threats to reliability with which we have to contend.

Two useful rules to remember when considering exhaust valves are:

1. An improvement in combustion efficiency, such as a change to a higher compression ratio (so long as this does not intro-duce a tendency to pre-ignition) will reduce exhaust valve temperatures.
2. an improvement in volumetric efficiency will raise exhaust valve temperatures.

Thus, if we live in Denver, Colorado, or Nairobi, Kenya, at a high altitude and the atmospheric pressure much lower than that at sea level, and we raise the compression ratio of our car to try to bring the power back to sea level horse power we will lower the working temperature of the exhaust valves and improve their reliability. If, on the other hand, we leave the compression ratio unchanged and fit a supercharger, exhaust valve temperature will rise.

Exhaust valve size

To the unwary it would seem an obvious tuning improvement to fit the largest exhaust valve that the combustion chamber will accommodate. The temptation to go to the limit must be tempered with a knowledge of the dangers involved. A large valve will get rid of the exhaust gases quickly. This cannot be denied, but the amount of heat picked up by the valve during combustion and exhaust varies as the square of the head diameter. Approximately 75 per cent of this heat must be rejected through the valve seat. The rest travels down the stem and out through the guide (see Fig. 81). The seat area varies directly as the valve head diameter.

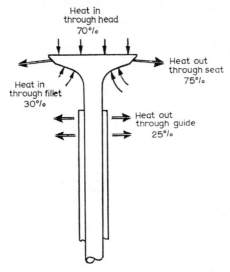

Heat in
through head
70%

Heat out
through seat
75%

Heat in
through fillet
30%

Heat out
through guide
25%

FIG. 81.   Heat balance on typical exhaust valve

Thus if we increase the head diameter by 50 per cent we increase the reception area for picking up heat by 125 per cent. The total heat rejection path, however, will only be increased by about 38 per cent. Any attempt to increase the heat path by the use of extra wide seats is doomed to failure, since a certain minimum closing pressure is needed to establish a good contact between the two seats and to bite through the film of carbon and lead compound that is deposited on the seats. Valve seat width then should be as wide as possible but can seldom exceed 0·100″ even on the larger 1·5″ to 1·75″ valves. Another danger from the use of too large exhaust valves is the increased inertia loading on the valve itself. The weakest part of the hot valve is the junction of the valve head and the stem. The use of a valve head that is too heavy can lead to 'stem-stretching' and the eventual loss of the head. Always beware of the engine that loses its exhaust valve clearance after a short period of running. It could be sinkage of the seat; but again it could be stem-stretching. A new set of valves in a material with a higher hot strength cost much less than a damaged head and piston.

The correct size for an exhaust valve is now well established between fairly narrow limits. Expressed in terms of the inlet valve dimensions, the exhaust valve throat diameter should be approximately $\frac{7}{8}$ of the inlet valve throat diameter. Any ratio less than $\frac{4}{5}$ should be regarded as inadequate for a high speed engine. Since the available space for the accommodation of valve heads in conventional cylinder heads must be shared out between

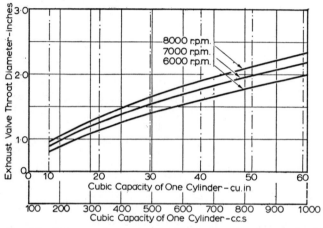

Fig. 82.   Exhaust valve throat sizes (Bath-tub and wedge type heads)

the inlet and exhaust valve heads there is nothing to be gained by making the exhaust valve bigger than necessary at the expense of the inlet valve. As a general rule then the tuner who cannot embark on an expensive programme of air-flow experiments to strike the right balance between inlet and exhaust valve sizes must be guided by our simple rule. The three curves of Fig. 82 are based on this ratio of 7:8 and the inlet valve throat sizes from Fig. 74.

EXHAUST VALVE THROAT AND PORT SHAPE

On the exhaust valve there is never any question of deliberately reducing the size of the throat. When a valve seat insert is used there must be some slight reduction in diameter below that of the inner seat diameter or the thickness of the insert will be insufficient to withstand the pressure of the interference fit. Apart from this limitation one should always make the exhaust valve throat as large as possible for the particular head size. The shape of the port, especially in the vicinity of the guide, is dictated by conflicting requirements:

    (*a*) good gas flow

    (*b*) good heat flow.

On a well designed exhaust valve approximately 25 per cent of the heat picked up by the valve head flows away down the valve

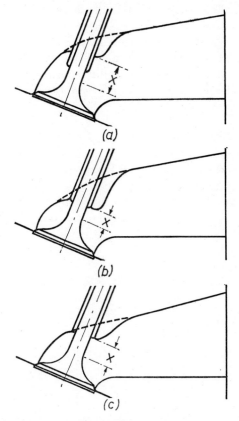

FIG. 83. Exhaust port shape

stem to the guide and then to the coolant. At (*a*) in Fig. 83 we show an exhaust valve port with a good unrestricted gas passage, but the heat flow path down the stem to the coolant is poor. The exposed guide will overheat. This exposed guide is a particularly bad feature. By picking up heat from the gas stream it will tend to act as a heat barrier between the stem and the metal of the cylinder wall. Example (*b*) is typical of good modern production practice. The stem diameter is well proportioned. For a small engine the stem diameter should be at least $\frac{9}{32}$ inch, for a medium $\frac{5}{16}$ inch and for large engines of 3 litre or more at least $\frac{11}{32}$ inch.

The portion of the stem exposed to the hot gases in (*b*) is kept short. The distance *x*, from the neck of the valve to the end of the guide, is very short. Unfortunately, the excellent heat flow given by design (*b*) is obtained at some sacrifice of gas flow. For a high performance engine the writer prefers design (*c*) in which the guide boss has been ground down to be almost non-existent and the end of the guide has been blended into the main port contour. From this point on the port is expanded at about 10-15 degrees included angle until it meets the manifold joint face.

Beyond the cylinder head

In this chapter we have confined ourselves to the related problems of getting the new gas in and letting the old gas out, but only within the confines of the cylinder head. The part played by the induction system has been covered in Chapter Four. The design of the exhaust system will be discussed in Chapter Eleven.

CHAPTER ELEVEN

Special Techniques

MULTI-CARBURETTORS

WHEN SUPERTUNING an engine we must always have a well-defined goal in mind. Nine times out of ten the amateur tuner starts with the vague aim of extracting more power just short of mechanical disaster. Such a policy often destroys the whole roadworthiness of the car. What use on the modern highway, especially in Britain, is a useful power curve that only begins at 4000 r.p.m. and makes you grab for the gear lever when the speed drops to 4500 r.p.m.?

Here is a story with a moral for all tuners. Back in the mud-plugging thirties a certain physically-handicapped West of England driver with an uncanny delicate touch with his hand-operated throttle and clutch controls was winning trials with a J2 M.G. Midget. Competition included the larger-engined TA Midget, some with large-bore carburettors and high-compression heads, some even with big superchargers. Our J2 driver, however, threw away his twin carburettors and fitted a small-bore single. In this way he was able to maintain a high torque at very low engine speeds. When making a standing start on a slippery surfaced hill it is low speed torque that counts. The stall point of the race-tuned TA Midgets was so high that the drivers often seemed to be digging their own graves with their wildly spinning wheels. The moral of the story is this: decide what you want, then tune to get it.

When making the decision on the number and size of carburettors to be used on our modified machine we should always think clearly about the future that is planned for the car. Will it be high-speed touring, rallies, speed hill climbs or circuit racing? These four purposes lie on an ascending scale of tuning stages, high speed touring being Stage One and circuit racing Stage Four.

One choke per cylinder

It is common knowledge in tuning circles that the use of one

carburettor (or choke) per cylinder opens up a wonderful area of
induction pipe tuning that is largely lost to us when two or more
cylinders share a carburettor. It will help us to our better understanding of the principles involved if we consider how the gas flows
in the two systems.

The more cylinders drawing off one carburettor the less will be
the variation in velocity through the venturi and the higher will be
the induction vacuum reading registered by a well-damped vacuum

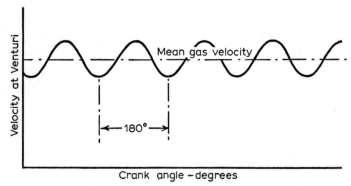

FIG. 84.   4 cylinders per carburettor

gauge. Fig. 84, although oversimplified, serves to illustrate the
manner in which the gas velocity fluctuates at the venturi of a
single carburettor feeding four cylinders. In practice more complex
wave-forms would be given. Fig. 85 shows the fundamental
simplified velocity wave that would occur in the venturi of a single

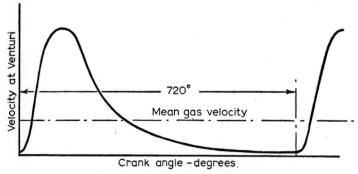

FIG. 85.   1 cylinder per carburettor

carburettor feeding one cylinder of an engine. Depending upon the length of the pipe and the speed of the engine a more complex wave-form would be superimposed on this fundamental wave, but the two simplified diagrams serve to illustrate the essential difference between the two systems that the gas flow is almost steady and continuous in one, surging widely in the other. By careful design one can make these surges work for us to push a large charge of mixture into the cylinder. This is called ram charging. The fewer cylinders per carburettor the more ram-charge we can get. The correct length of induction pipe to give maximum ram effect is called 'the tuned length'. This ram effect is so much a part of the one-choke-per-cylinder technique that a little time must be devoted to a discussion of the phenomena involved.

Let us consider the simple case of a single-cylinder 4-stroke engine fitted with a long inlet pipe. The pipe is supposed to be of constant cross-section throughout—no bends and no restriction at the carburettor venturi. When the inlet valve opens near T.D.C. and the piston begins to descend a negative pressure pulse is made

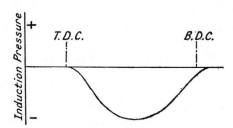

Fig. 86.   Variation in inlet port pressure
during induction stroke

to travel down the inlet pipe towards the open end. It travels at the velocity of sound in air, i.e. at approximately 1100 ft. per sec. If we were to fit a pressure pick-up behind the valve and feed the signal to an oscilloscope and if we could eliminate interference from all other wave forms present at the same time, we would obtain the simple trace of the form shown in Fig. 86. When this negative pulse reaches the open end of the pipe it will produce a rarefaction in the atmosphere near the end of the pipe. The surrounding air will rush in to fill this depression and by virtue of its inertia will produce a *positive* pulse that will travel back up the pipe with the speed of sound in air. This is called a 'reflected pulse', the first reflection. This reflected pulse, when it reaches the cylinder, is reflected again back towards the open end of the pipe. As our pressure pick-up is connected to the inlet pipe near the valve head, both arriving and receding pulses will be recorded almost simultaneously. As they are of the same sign, i.e. both positive, they are effectively the same pulse. When this positive pulse reaches the end of the open pipe it again is reflected as it expands into the atmosphere, this time as a *negative* reflection. This is the second reflection. We would therefore see in our instrument a sequence of negative, positive, negative, positive pulses, each one

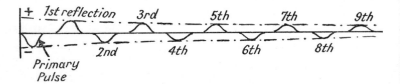

Fig. 87. Approximate attenuation of a single induction pulse during nine reflections

slightly smaller than its predecessor; this sequence continuing, if permitted, until the pulse eventually died away. This, of course, is not permitted to happen, since another induction pulse is soon born and this gets mixed up with our existing pressure wave in the manner we will show later. For simplicity we have shown in Fig. 87 what the sequence of pulses would look like on the oscilloscope if this interference from subsequent induction processes did not take place. There is a slight reduction in the magnitude of the pulse at each reflection, this reduction depending upon the shape of the valve port and the shape of the carburettor entry. A fair average for this reduction would be 12 per cent. From this we would expect a sequence of reflections of the following amplitudes:

Original pulse	1·00
1st reflection	0·88
2nd reflection	0·77
3rd reflection	0·68
4th reflection	0·60
5th reflection	0·53

etc.

The pulse dies out so slowly that residual ripples will exist in the pipe when the inlet valve opens again two revolutions later.

The time in seconds taken for the pulse to travel down the pipe and back again is

$$T = \frac{L}{6V_s} \tag{12}$$

where L = the pipe length in inches
V_s = the velocity of sound in air at the temperature and pressure in the induction pipe, about 1100 ft. per sec.

If we let N be the engine speed in r.p.m., the time interval can be converted to crank-angle degrees:

$$t = \frac{NL}{V_s} \tag{13}$$

Unless the pipe is a very long one, the pulses will overlap, something in the manner of Fig. 88. The separate pulses are shown as dotted lines, the resultant wave as a full line. The resultant wave is simply the ordinates of the overlapping pulses added algebraically, a positive ordinate of 2·5 lb. per sq. in. coming on top of a negative ordinate of 1·3 lb. per sq. in. would yield a resultant of 1·2 lb. per sq. in. positive. The shape of the original pulse and the subsequent

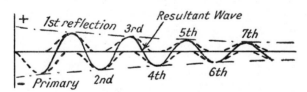

FIG. 88. The resultant wave

reflections will be roughly that of a sine curve, since the piston movement is approximately sinusoidal. The half-period of the wave or pulse will however be slightly more than 180 degrees of crank-angle movement since at the end of the pulse the piston has moved a stroke-length away further from the valve port. This gives the effect of a lag at the end of the pulse. This lag will amount to about 10 crank-angle degrees at an engine speed of 5000 r.p.m.

(see Fig. 89). The amplitude of the pulse will increase roughly as
the square of the engine speed, being about $3\frac{1}{2}$ times greater at
5000 r.p.m. than at 2500 r.p.m.

Now, let us consider what is happening at the end of the induction
stroke when the inlet valve is closing at, say, 50° after B.D.C. Very
little gas flow can occur, however, during the last 30° of valve

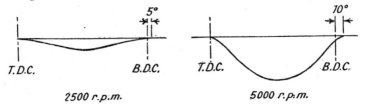

FIG. 89.   Effect of engine speed on amplitude of primary pulse

closing and it is more realistic for our purpose to consider the
effective inlet valve closing (E.I.C. in Fig. 90) as taking place about
20° after B.D.C. The hypothetical curve in Fig. 90 is constructed
to show typical values of the pressures existing in the cylinder and
in the valve throat when $t = 90°$. The higher the cylinder pressure
curve (curve B) at the point of E.I.C. the higher will be the volumetric efficiency and the higher the engine power. It would appear
from this that the optimum volumetric efficiency will be given when

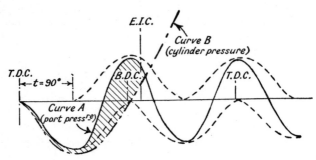

FIG. 90.   Inlet port pressure before effective inlet valve closing point: $t = 90°$

the pressure in the valve throat is as high as possible at E.I.C.
Moreover, it is the area between the two curves that influences the
passage of gas into the cylinder during the critical $30 - 50°$ before
E.I.C. This area is indicated by shading in Figs. 90, 91 and 92.
Figs. 91 and 92 show the cylinder and valve throat pressures when
$t = 60°$ and $t = 120°$ respectively.

For a high cylinder pressure at E.I.C. two conditions must be met:

(a) curve A must be as high as possible at E.I.C.

(b) curve A must maintain a much higher pressure than curve B for a period of about 60° before E.I.C.

An examination of the three cases shows that these two requirements tend to work against one another. In Fig. 91 the port pressure

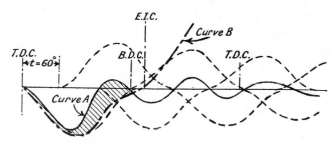

FIG. 91. Inlet port pressure before effective inlet valve closing point: $t = 60°$

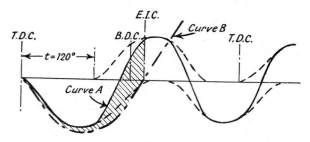

FIG. 92. Inlet port pressure before effective inlet valve closing point: $t = 120°$

at E.I.C. is low and volumetric efficiency will also be low. In Fig. 92 port pressure is high at E.I.C. but the difference in pressure between curve A and curve B over the critical 60° before E.I.C. is not at an optimum. An optimum would appear to occur around the value $t = 90°$.

Experimental measurements

About twenty-five years ago the writer was privileged to assist in the analysis of the results of experiments on two sizes of single-

SCE Q

cylinder engine in which power was measured with different lengths of induction pipe. War-time secrecy prevented their publication at the time. The results can be summarised in one sentence. The optimum value of t, over a range of engine speeds from 2000 to 3600 r.p.m., was found to vary from 80 to 89°. The value of t in each case was not very critical and if a value of 85° were chosen for each speed the power developed would only be 2 or 3° below optimum in the worst cases.

Recommended pipe lengths

We must emphasise that the above results are based on two engines and only cover what is to us the middle speed range. The

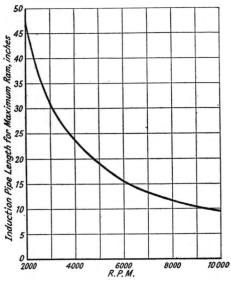

FIG. 93.   Optimum induction pipe length for different engine speeds. The effective pipe length is measured from the valve head

writer, however, has used the design data from these tests for many years and when circumstances have prevented him from using a dynamometer to tune the induction system he has used the curve of Fig. 93, based on a value of $t = 85°$, to determine the correct induction pipe length. It is not often that the results are found to be in error.

When designing an induction system for an unsupercharged

racing engine or a competition sports car engine it is customary to design for maximum ram to occur at maximum power r.p.m. This boosts the maximum speed on the straight and by use of the gearbox the driver can minimise the falling off in power at the lower speeds where the ram is poor. The influence of ram is much greater at the higher speeds than at low speeds (see Fig. 89). Thus an induction system tuned to give a gain of 10 per cent at 6000 r.p.m. will only lose about 3 per cent of the power at 3000 r.p.m. There are times when a tuner fits longer trumpets or air-horns to his carburettors to lower the point in the power curve where ram occurs. He would do this for a particularly twisty circuit where acceleration is more important than top speed.

Ideally, there should be no type of restriction, bend, or ex-crescence in the induction pipe for the maximum ramming effect to be obtained. Thus a twin o.h.c. engine with only a slight curve after the valve throat is very close to the ideal. The slight constric-tion of the carburettor venturi introduces reflected waves of small magnitude that conflict with the simple wave form of our theory, but the effect is not usually found to be of consequence. Where fuel injection is used as on the D Type Jaguar and the 300 SL Mercedes Benz the tuned length of induction pipe can be of con-stant cross-section all the way from air intake trumpet to inlet valve.

Carburation can sometimes be a problem with tuned induction systems. A pulsating air flow across the main jet induces a greater fuel flow than would be given by a steady flow. Thus a tendency for an overenrichment will occur at the part of the power curve where ram appears. A second problem is the spray of fuel that the outwardly travelling pulse ejects from the open end of the car-burettor intake. When Weber carburettors are used special venturi extensions can be fitted to reduce this loss. These extensions take the form of cruciform baffles that extend between the auxiliary venturi and the mid-point of the air-horn. Although described as air-straighteners in the Weber Technical Handbook the main purpose of these baffles is to catch the fuel sprayed out by ram pulsations.

A good air cleaner will also absorb this spray, but not everyone is prepared to sacrifice the slight loss in power that inevitably occurs when air cleaners are fitted.

PETROL INJECTION

Petrol injection was very much in the news about eight years ago. At that time Detroit was busy planning the introduction of five or six systems as optional equipment for the 1957 models. Sales predictions were too optimistic, however. Sales were very poor and in the language of Detroit 'injection was a lemon'. To-day it survives as an efficient but expensive option on America's only real sports car—the Corvette. In Europe direct cylinder injection, in the manner of the Diesel engine, is still seen on the standard 300 SL Mercedes Benz. Port injection using the Lucas system was a competition success on the D type Jaguar and has recently been adopted by Maserati for one of their GT models.

Port injection systems

These can be sub-divided into

(*a*) continuous injection, and

(*b*) timed injection

Continuous injection

Hilborn-Travers. This system has been in use for many years at

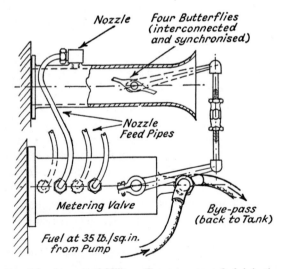

Fig. 94. Layout of Hilborn-Travers patent fuel injection system (four-cylinder engine)

Indianapolis and on Dirt Track Midgets. It is largely an 'On-Off' device and is not bothered much with the subtleties of metering at part throttle. A simple metering valve controls the output of a vane type fuel pump driven by the engine and distributes the fuel to simple injectors in the manner of Fig. 94. For a sports car the Hilborn-Travers system has been found to lack flexibility. For its designed purpose the system is excellent, but for the wide range of operating conditions demanded by a road car a more sophisticated metering system is demanded.

The Rochester Continuous Injection System. There is nothing simple about the Rochester system, but many years of development work have made it thoroughly dependable. It is available as optional equipment on the Chevrolet Corvette sports car. The 1962 327 cu. in. Corvette engine with large 4-barrel carburettor, a 'three-quarter race cam' and a compression ratio of 11·25 is rated at 340 S.A.E. horse power. The substitution of the Rochester fuel injection system increases the power by 20 horse power—an increase of about 6 per cent.

The Rochester system uses the slight pressure drop across a large venturi placed at the inlet to the system to signal the control system on the fuel metering device. Fig. 95 shows the essentials of the fuel

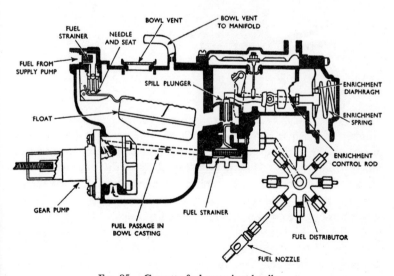

FIG. 95. Corvette fuel metering details

metering system. A conventional automobile diaphragm pump
supplies fuel to the float bowl as in the carburettor system. In the
base of the float bowl is an engine driven pump that draws fuel
from the bowl and supplies it at high pressure to the metering
system. Flow to the nozzles, positioned to inject downstream into
the eight individual ports, is regulated by the position of the spill
plunger (Fig. 95). This position is controlled by the pressure
exerted by the fuel control lever (see Fig. 96). Operation of the

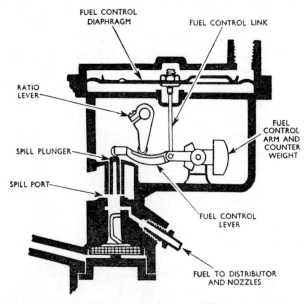

FIG. 96.   Corvette fuel control linkage

control mechanism is as follows. Pump pressure acts under the spill
plunger to push it upwards against the fuel control lever. This
pivots on the ratio lever to transmit a downward pull to the fuel
control diaphragm. The vacuum signal from the venturi acts in
the opposite direction. Fuel metering is thus achieved by a balance
between these two forces. When the spill plunger is in a high
position more of the pump output is spilled back to the float bowl.
When it is in a low position a greater percentage of the total pump
output is fed to the injectors. The position of the ratio lever varies
the leverage between the spill plunger and the diaphragm. By
connecting the ratio lever to the manifold vacuum provision is

made for an automatic enrichment of the mixture at full power.

The idle system is shown in Fig. 97. During idle operation when the throttle valve is closed air enters the manifold through a by-pass. A large idle air adjustment screw is used for idle mixture tuning. This operation should be carried out exactly as described for the

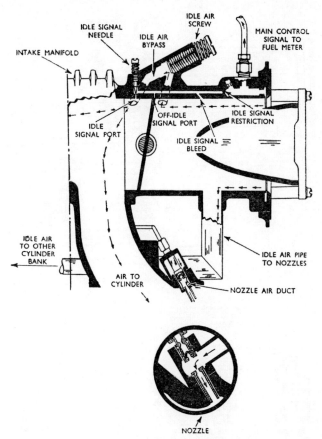

FIG. 97.   Main control signal and idle operation

setting of the idling mixture on a typical carburettor. A separate air duct is also taken to each nozzle. This serves to 'break suction' on the nozzle orifice. Without this provision changes in manifold vacuum would have a pronounced influence on fuel metering.

The position of the enrichment lever and the ratio lever is

controlled by manifold vacuum acting on the spring-loaded enrichment diaphragm (shown to the left of Fig. 99). The two mixture limit stops that restrict the travel of this diaphragm are calibrated at the factory and should not normally be re-set during subsequent tuning operations.

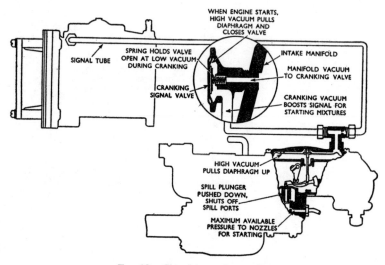

FIG. 98.   Signal boost for starting

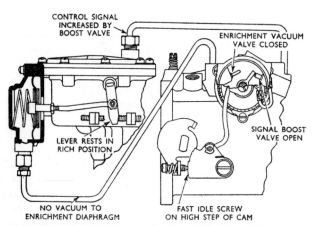

FIG. 99.   Cold enrichment, first stage signal boost plus power enrichment

To supply a very rich mixture for starting a vacuum valve is opened to admit manifold vacuum to the top of the main control diaphragm (see Fig. 98). Since this vacuum is very large in comparison with the vacuum signal from the venturi, full rich mixture is provided. The inset in Fig. 98 shows how this vacuum can only reach the main control diaphragm when the vacuum 'pulled' by the engine is the lower vacuum of 'cranking speed', i.e. no more than 12 inches Hg. As soon as the engine starts and a higher vacuum is pulled the little diaphragm is pulled to the right to seal off the connection between the manifold and the main control diaphragm. The cold-running arrangements after starting are shown in Fig. 99. The first action is to cut off the vacuum to the enrichment diaphragm. This puts the ratio lever into the rich position. The second action is the opening of the bleed valve to introduce an increased vacuum to the passages that transmits the signal to the main control diaphragm. A thermostatic control provides a gradual relaxation of the mixture enrichment as the engine warms up to temperature.

Timed Injection

The Lucas System. The Lucas system of petrol injection was fitted to the Works D Type Jaguars that raced so successfully at Sebring, Nurburgring, Rheims and Le Mans in 1956. A recent application

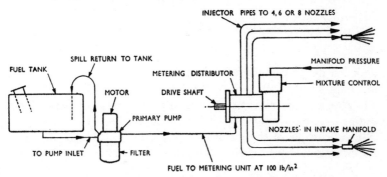

FIG. 100. Arrangement of Lucas petrol injection system

has been to a GT Maserati. The system is shown schematically in Fig. 100. The motor driven fuel pump is mounted on the chassis near the fuel tank. This feeds fuel to the distributor which is driven by a suitable auxiliary drive on the engine. The function of the distributor and its mixture control is to meter precise quantities of fuel to each cylinder in turn at the optimum time during the induction stroke as indicated by dynamometer tests on the particular installation. The fuel nozzles are located in the induction manifold near the inlet ports. Good mixing and evaporation of the fuel during the intake and compression strokes can only be achieved by a certain amount of experimentation on the dynamometer. The

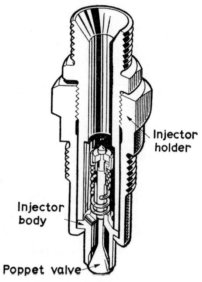

Injector
holder

Injector
body

Poppet valve

FIG. 101.   Lucas injector

designed delivery pressure of the primary pump is 100 lb. per sq. in. Excess fuel is returned to the fuel tank through a relief valve. The choice of atomiser again is usually the subject of work on the test-bed. A typical injector unit is shown in Fig. 101.

The principle of the metering distributor is shown in Fig. 102. As the engine driven rotor rotates, first the port at one end is uncovered to admit fuel under pressure and the port at the other end connects the rotor chamber to one of the fuel injection lines. Further rotation uncovers new ports and what was previously the delivery end of the rotor receives fresh fuel under pressure from the pump; the opposite end is now in communication with the next injection pipe to be charged with fuel. As the shuttle is subjected to

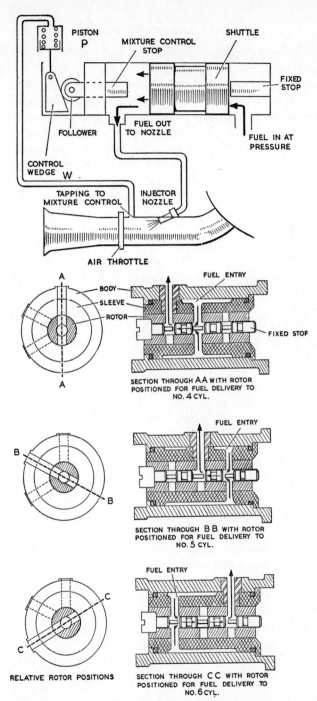

PISTON
P

MIXTURE CONTROL
STOP

SHUTTLE

FIXED
STOP

FUEL OUT
TO NOZZLE

FUEL IN AT
PRESSURE

FOLLOWER

CONTROL
WEDGE W

TAPPING TO
MIXTURE CONTROL

INJECTOR
NOZZLE

AIR THROTTLE

A

A

BODY
SLEEVE
ROTOR

FUEL ENTRY

FIXED STOP

SECTION THROUGH AA WITH ROTOR
POSITIONED FOR FUEL DELIVERY TO
NO. 4 CYL.

B

B

FUEL ENTRY

SECTION THROUGH BB WITH ROTOR
POSITIONED FOR FUEL DELIVERY TO
NO. 5 CYL.

C

C

FUEL ENTRY

RELATIVE ROTOR POSITIONS

SECTION THROUGH CC WITH ROTOR
POSITIONED FOR FUEL DELIVERY TO
NO. 6 CYL.

Fig. 102. The basic elements of the Lucas injection system

pump pressure on alternate sides, it reciprocates between two stops, the fixed stop and the mixture stop. At each stroke of the shuttle a chamber full of fuel is injected into an injection line. The quantity of fuel injected depends upon the stroke of the shuttle, which is controlled by the position of the variable mixture stop. This stop is moved in and out under the action of a piston-operated wedge, high induction vacuum shortening the stroke, low vacuum lengthening it. This wedge is shown schematically in Fig. 102 and in correct detail on the twin-rotor D Type Jaguar distributor in Fig. 103. The two rotors shown in this model are driven at one-

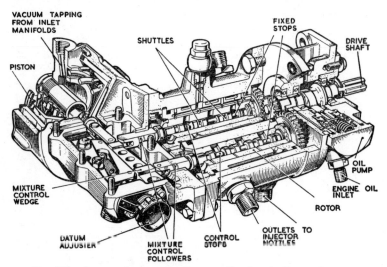

FIG. 103.   Lucas twin-rotor distributor as fitted to the D type Jaguar
(*'The Autocar'*)

quarter engine speed. Each rotor contains two shuttles. Although more complicated than the simple system described in Fig. 102 the operating principle is identical. A datum adjuster is also provided to permit tuning adjustments before each race to suit the particular atmospheric temperature, pressure and humidity conditions.

On the Jaguar engine a sliding plate type of throttle is used and at full-throttle a completely unobstructed intake pipe is provided to each cylinder. Fuel injection on the D Type Jaguar is made into each separate intake branch about four inches ahead of the cylinder head flange. Injection is against the airstream. Other earlier

Lucas installations gave good results with the fuel injected with the airstream. A characteristic of the D Type Jaguar is its ability to run down to 10 m.p.h. in top gear. This excellent low-speed torque in an engine with such a high specific power output can only be credited to the fine atomisation and good mixing given by the fuel injection system.

Direct Cylinder Injection

Such a system must of course be timed if it is to work at all. The German Bosch system has been used on the 300 SL Mercedes Benz sports car since its inception. In many ways the system resembles the fuel injection used on modern Diesel engines. There are two essential differences. First the air supply is throttled. Diesels invariably run with an unthrottled air intake. Second, control of fuel metering, as in the Rochester system, is based on a vacuum signal from a venturi placed at the entry to the induction system.

The Bosch pump, in common with the pumps used on most European Diesel engines, has a separate cam-operated plunger for each cylinder with the output from each plunger controlled by helical ports in the annular sleeves. Rotation of the sleeves to increase or decrease the quantity of fuel injected at each upward stroke is performed by a rack which engages teeth machined on the outside of each sleeve. It is in the regulation of the movement of this rack that the petrol injection pump differs from the oil engine pump. An oil engine usually runs all the time at full throttle (in other words no throttle is provided) and power output is controlled entirely by fuel metering. A petrol engine, however, is fairly critical in its air/fuel ratio requirements and any attempt to control power output by using a wide-open throttle and cutting down on the metered amount of fuel would, at low powers, call for an air/fuel ratio so weak that the engine would fail to burn it. 'Stratified-charge' petrol engines that are capable of extending the weak range of burning to mixtures as weak as 50 to 1 are still in the experimental stage, and need not be discussed here. In the Bosch petrol injection system the position of the rack, and consequently the amount of fuel injected per stroke, is dictated by two main controls. A large diameter venturi is provided at the inlet to the air cleaner. The slight depression created by this venturi is used to operate a diaphragm which is connected to the end of the rack

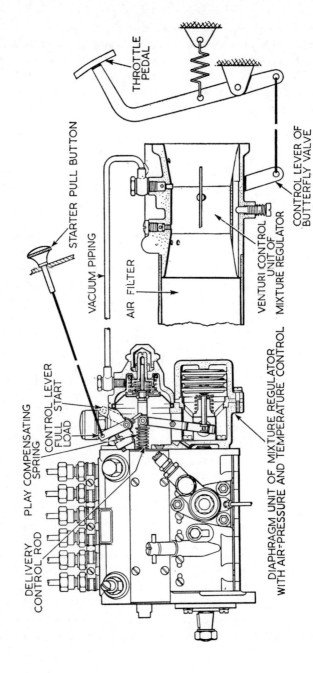

Fig. 104. The basic elements of the mixture controls on the Bosch injection system. (*The Autocar*)

see (Fig. 104). The venturi is thus a simple air meter and fuel is metered to match this air flow. An overriding control in the form of an aneroid capsule, connected to the rack by a rocking lever, compensates for variations in air temperature and pressure.

The injectors spray direct into the combustion chamber, injection occurring during the early part of the induction stroke and proceeding over a period of about 60 crank-angle degrees. With a timed injection system directly into the cylinder large valve overlaps can be used with no danger of fuel being lost through the open exhaust valve, since fuel injection need not commence until after the exhaust valve has closed. The Mercedes Benz engineers claim that fuel injection gave an increase in power output of 15 per cent over the carburetted engine. They do not state however how many carburettors, or of what size or make, were used in the comparative tests.

The 300 SL is notable for its smooth power and wonderful flexibility. One can only regret that the expense of direct fuel injection must limit its application to cars in the upper price bracket.

THE INTAKE AIR

The air intake on a normal road vehicle should always be filtered. A simple steel or copper wool filter is better than nothing, but the modern treated paper filters are extremely efficient and reduce power output by a negligible amount if changed at the recommended mileages. Modern competition sports cars usually operate without any air-filtration in order to extract the last horse power from the engine. It is therefore a matter of some importance to choose the position of the external air intake to draw air from a zone that is not heavily-laden with dust disturbed by the passage of the particular car or by dust thrown up by the car in front. Experience has shown that the top of the bonnet is a good compromise, since it is well above the road surface and is also well suited to the other requirements of the air-intake on the modern competition sports car.

Cold air intakes

Since motor-racing is in general a warm weather sport there is everything to be gained by trying to draw the carburetted air from as cool a source as possible. Under-bonnet temperatures as high as 180°F (82°C) are not uncommon in America when a car is left to stand in the sun. When the car is in motion the under-bonnet temperature can still be much higher than the external air temperature.

If we take an under-bonnet temperature of 50°C, with an atmospheric temperature of 20°C, the gain can be calculated from Charles' Law:

$$\frac{\rho_c}{\rho_h} = \frac{273 + t_h}{273 + t_c}$$

where ρ_c = the density of intake air with external intake

ρ_h = the density of intake air with under-bonnet intake

t_c = atmospheric temperature, °C

t_h = under-bonnet temperature, °C

$$\therefore \rho_c = \frac{273 + 50}{273 + 20} \times \rho_h$$

$$= 1 \cdot 10 \rho_h$$

This represents a gain of 10 per cent in air density, and an identical gain in indicated horse power. The gain in brake horse power will be approximately 8 per cent, or 1 per cent for every 4 degrees drop in aspirated air temperature. When the atmospheric temperature drops within 10°C (18°F) of Freezing Point a warm air intake is to be recommended since the danger of carburettor icing is at its greatest just above freezing point.

Forward ram intakes

A forward ram intake is sometimes used to give a slight boost to the induction system from the forward velocity of the car. With a forward ram intake the forward-facing air scoop must lead into a sealed duct which enters an air-box built around the carburettor air-horns. This air-box is usually made large enough to convert most of the kinetic energy into pressure energy before the air reaches the air-horns. To prevent any serious upset of the normal mixtures fed to the carburettors it is essential that balance pipes be led from the air-box to sealed covers on the carburettor float bowls.

Theoretically one could recover the whole of the velocity head of the airstream striking the front of the car, but in practice, since the airstream must be directed round a right angle to enter the carburettors and must give up some energy in overcoming frictional resistance in the duct, the most we can hope to recover is about 90 per cent of the total velocity head.

This velocity head is

$$p = \frac{\rho\, v^2}{288g}$$

where p = the velocity head, lb. per sq. in.
ρ = the air density, lb. per cu. ft.
v = the car velocity, ft. per sec.
g = the gravitational acceleration, ft. per sec.^{2}

If the car is travelling at 150 m.p.h. (220 ft. per sec.)

$$p = \frac{0 \cdot 076 \times 220^2}{288 \times 32 \cdot 2}$$

$$= 0 \cdot 4 \text{ lb. per sq. in.}$$

This represents a gain in power of about 3 per cent.

Only the really high speed sports car can gain much from a forward ram intake. At 75 m.p.h. the ram would be reduced to a quarter of the above value and the gain in power would be only 0·7 per cent. The nett gain must be even less, since the air scoop increases the drag of the body.

RAMMING EXHAUST PIPES

When a straight-through silencer is used it is possible to obtain a useful degree of ram-effect from the exhaust system to assist in the induction of an even higher mixture charge. To obtain the full effect, however, a separate exhaust pipe should be provided to each cylinder. Apart from sprint and drag-race machines one seldom finds that such a cumbersome arrangement can be tolerated. The Grand Prix B.R.M., at the beginning of the 1962 season, was an interesting exception. Before discussing the possibility of ram-charge on manifolded exhaust systems we must give a brief description of the working principle on the single exhaust stack.

The pressure fluctuations in an exhaust pipe are built up of an original pulse and subsequent reflected pulses exactly as described earlier for the induction system. In this case, however, the original pulse is positive. The correct length of pipe to give the maximum assistance to the induction process is that length which makes the peak of the first reflection coincide with T.D.C. This is shown in Fig. 105. The aim is to extract as much as possible of the residual gases during the overlap between the opening of the inlet valve and the closing of the exhaust valve, a period of about 25 degrees with conservative valve timings and as much as 80 degrees with competition camshafts. The effect of this 'last-minute suction' is twofold.

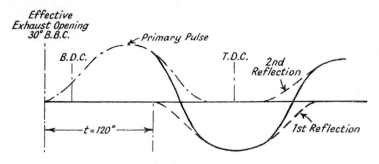

FIG. 105.   Exhaust port pressures near T.D.C.

In the first place it extracts some of the hot residual gas, thus helping to lower the final charge temperature. Secondly, by lowering the pressure in the cylinder it encourages the initial flow of the charge through the inlet valve. With touring-type overlaps an increase in volumetric efficiency of from 5 to 7 per cent can be obtained in this way. If we are willing to sacrifice fuel economy and a smooth idling speed, greater gains are possible by using overlaps of 50, 60 or even 80 degrees. In this way, at the speed of maximum ram, the residual gases are scavenged more effectively at the expense of some petrol mixture going to waste down the exhaust pipe. The cooling effect of the escaping mixture on the hot exhaust valve is an added benefit, this effect being much more pronounced when alcohol fuel, with its very high latent heat, is used. Obviously the power produced by the engine will be reduced below normal at those engine speeds that produce a positive pressure in the exhaust port during overlap. More than normal residual gas will be retained and less than normal fresh charge will be induced. At idling speeds the abnormally early opening of the inlet valve results in a back-flow of exhaust gas into the inlet port.

From Fig. 105 it is seen that the maximum depression at T.D.C.

is given when $t = 120°$. In calculating the value of L, the pipe length in inches, from the formula $t = \dfrac{NL}{V_s}$, the value of V_s to be used is larger than that used for the induction system. The velocity of propagation of the pressure pulses varies inversely as the square root of the density of the gas. Not only is the exhaust gas much hotter than the mixture in the induction system, being about 800°C at the port and about 150-300°C at the tail-pipe, but the gas

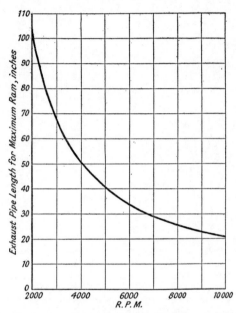

FIG. 106.   Optimum exhaust pipe length for different engine speeds. The effective pipe length is measured from the valve head

contains substantial amounts of water vapour which is lower in density than air. A fairly representative value of V_s for a sports car exhaust system at maximum power would be 1700 ft. per sec. Taking the above values the curve in Fig. 106 has been constructed to give an approximation to the optimum exhaust pipe lengths for a range of engine speeds.

Branched exhaust pipes

Multi-branch exhaust systems can be tuned on the dynamometer

to find the optimum length. In general the application of theory
to the problem is complicated by the presence of residual waves
from the previous cycle, plus the overlapping wave systems from
the 4, 6 or 8 cylinders exhausting into the common system. While
we cannot accurately predict the exact pipe length for maximum
ram at any chosen engine speed we can at least give some guidance
in indicating the best manifold shape to take full advantage of the
extractive effects. Fig. 107 illustrates two good systems and one bad

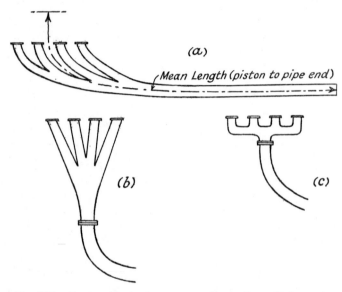

Fig. 107. Good and bad exhaust systems for the four-cylinder engine

one. No useful extractive effort could be expected from system (c).
Not only is the system inefficient for the steady flow of gases, but
the effect of several branches leading at right-angles into a common
manifold is to subject the exhaust process from individual cylinders
to harmful interference from the pressure waves in adjacent cylinders. Experience has shown that the extractive effect can be enhanced most by a system in which the branches enter the main
pipe at different points, as in (a). Each branch should enter at as
small an included angle as possible and the lengths of the branches
should be made as great as possible inside the limits set by body
and chassis configuration. System (b) is usually less difficult to

accommodate than (a), but every effort should be made to keep the branches as long as possible.

NEW IGNITION TECHNIQUES

Only conventional ignition systems have been considered so far, but the coil ignition system that has been with us so long with no changes of great note in the last thirty years or more, is now undergoing a metamorphosis. At the time of writing it is impossible to predict what will be 'the conventional system' in ten years time.

Transistorised systems

Transistorised ignition kits have been available in America for more than a year and reports from sports car owners all over

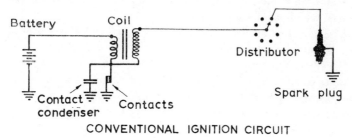

CONVENTIONAL IGNITION CIRCUIT

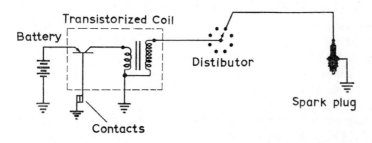

TRANSISTORIZED HIGH VOLTAGE IGNITION SYSTEM

Fig. 108. Schematic diagrams of conventional and transistor-switched high-voltage ignition systems

America confirm that more reliable high-speed sparking can be given by this device. The simple transistorised system replaces the condenser with a transistor as shown in Fig. 108. In this application the transistor is simply a relay switch *with no moving parts*. A very small current, about 0·2-0·3 amps through the normal contacts,

passed into the germanium crystals of the transistor switches a
current of 6-7 amps through the primary windings of the coil. No
condenser is required and the life of the contacts is extended. Plug
fouling tendencies are also reduced.

Transistorised kits

A typical American kit that has been extensively tested by
tuning establishments across America is the AEC77 made by
Automotive Electronics Co., 80 Wall St., New York 5, N.Y. This
is a negative earth kit. For British sports cars with positive earths
the AEC78 should be fitted. Each kit comprises three components:
a transistor unit mounted inside a finned heat sink assembly, a
special high voltage coil with a turns ratio of 1:300 and a one-ohm
ballast which is only used on 12 volt systems.

The heat sink assembly should be mounted in a fairly cool place
and should be well earthed to the car chassis. The old coil is
replaced by the new one and the condenser discarded. Plug gaps
can be opened by 0·005″ to 0·010″ above maker's specification.
Dwell angle can be set 2 or 3 degrees higher than normal. Plug
leads should be checked for possible leakage paths and the ignition
system tune checked in the normal way. Only one disadvantage
has shown up so far. When a transistor fails it fails completely and
without warning. If this happened far from home a replacement
transistor might be difficult to find. The simplest expedient would
be to refit the old coil and condenser. Transistorised kits are also
available from:

Auto-Marine Laboratories, Inc.,

6 East Main Street, Ramsey, New Jersey.

and

Electronic Ignition Corp. of America,

412 East Washington, Phoenix, Arizona.

A second stage in the application of modern electronics to the old
ignition system is that developed by the Tung-Sol Corporation of
America which uses a normal contact-breaker to trigger a thyratron
valve ('tube' in America). This works like a high voltage make-
and-break device. The thyratron discharges a capacitor through
the coil primary. The system is very effective, but expensive.

Capacitors are used in several new devices that will soon be on
the market. In these systems the battery provides power to an

oscillator, which in turn provides an alternating voltage of from 300 to 4000 volts, depending upon the particular design. This is rectified and stored in a large capacitor. In some systems there is no need for a coil, the discharge from the capacitor being passed

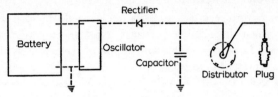

Fig. 109. Smits capacitor discharge system

direct to the plugs; in others it passes to the coil primary. One of the early systems, developed by Smits in Holland, is shown in Fig. 109.

Electro-magnetic systems

Delco-Remy and Motorola in America and Lucas in Great Britain have all developed new systems which stem generically from the magneto. In these systems small magnetic impulses, created by an engine-driven rotary device, are amplified and then discharged through the plugs. The American systems use an iron vane rotating inside a magnet to supply the initial signals. These signals are amplified by transistors, then passed to the coil primary.

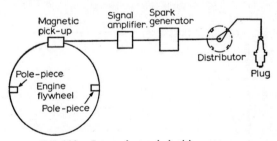

Fig. 110. Lucas electronic ignition system.

The Lucas system, now used on the B.R.M. V-8 Formula I engine, uses pole-pieces attached to the flywheel to generate small impulses in a magnetic pick-up as shown in Fig. 110. These impulses are magnified in a transformer and these higher voltage signals are passed to a transistorised spark generator. The system provides a

20,000 voltage secondary with a cycling time of less than 200 micro-seconds. A normal distributor is used. A disadvantage of the system is the high current consumption.

The piezo-electric system

Certain natural crystals, such as quartz, will develop a voltage potential when subjected to pressure. New developments in solid-state physics have led to the manufacture of ceramic crystals, based on lead-zirconate-titanate materials (usually abbreviated to PZT) that will generate 20,000 volts when subjected compressively to 7000 lb. per sq. in. This pressure is well within the safe stress limit of the material.

The Clevite Corporation of Cleveland, Ohio, use two PZT elements ¾ in. long by ⅜ in. diameter in their 'Spark-Pump'. These synthetic crystals are subjected to a load of 1200 lb. through a lever with a 15:1 leverage, the lever being cam operated from an engine auxiliary drive. This unit is now developed to the stage of ex-perimental production and is undergoing field trials on Go-Cart engines. The weight of the unit is 8 ounces and it occupies a space of 3½ cubic inches.

The chief attraction of this system is the extremely rapid voltage rise—about 10 millimicroseconds to reach 10,000 volts, hundreds of times faster than the fastest magneto ignition. The electrical

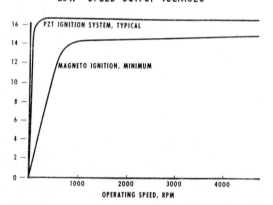

LOW SPEED OUTPUT VOLTAGES

FIG. 111.   Comparison of starting voltages: Magneto versus 'Spark-Pump'

leakage path presented by a badly fouled plug becomes almost non-existent with such rapid voltage rises. Starting, which has always been a weakness of magneto ignition, is excellent with the Spark-Pump. Fig. 111 shows that a voltage of 14,000 volts is developed at cranking speeds of less than 100 r.p.m.

One can see that such a system holds great promise as a racing engine development. Unfortunately multi-cylinder units are not yet available.

CHAPTER TWELVE

Cams and Blowers

SPECIAL CAMSHAFTS

ONE CAN do almost anything with cams, but no one has yet designed one that is ideal for every running condition of an engine. One can choose a profile that will give a smooth low-speed idle, a good fuel consumption and a torque curve that gives useful torque from 1000 r.p.m. upwards. At the other extreme one can grind a cam that lifts the peak r.p.m. to unimagined heights, that gives a wonderful surge of power so long as the speed never falls below 4000 r.p.m., that simply refuses to idle well at anything less than 1500 r.p.m.—and one can compromise between these two extremes, which is what the designer of a mass-appeal sports car must always do. The writer once worked on a single-cylinder research engine that had sliding camshafts that gave a choice of three different cam profiles on the inlet side and three different cams on the exhaust side. While the engine was running one could 'change gear' on the camshafts by moving a lever, converting the engine from a mouse to a tiger in a split second. Perhaps, some day, an ingenious designer will apply this principle to a multi-cylinder engine. Meanwhile we are stuck with our fixed camshaft.

In Chapter Ten some consideration was given to valve timing and overlap, but the practical aspects of camshaft and valve train design were not discussed.

Factory options and re-grinds

Many manufacturers of sports cars now list special camshafts. In general, car manufacturers tend to be slightly conservative. For real 'wild' cam-grinds one must go to the specialist grinders such as Ed. Iskenderian.

The cam profile

One still sees mention in textbooks of 3-arc cams (the cam nose was one arc, the flanks the other two). Such cam profiles are now

254

ancient history. Modern cams are generally of the multi-sine-wave pattern and the profile is usually resolved by mathematical computer to fulfil the more exacting requirements of the modern designer. The mathematical laws governing the design of modern cams are too complex and too specialised for a book of this character, but a general description of the problems involved will be appropriate. The modern designer recognises that the valve train is flexible, especially on the push-rod operated designs. He no longer chooses to ignore this fact when designing his cam profile. Due allowance is made for the initial storage of energy in the valve train and the release of this energy during the deceleration phase of the opening operation. The action of a cam in opening a valve may be considered in terms of four intervals. These are illustrated in Fig. 112.

Fig. 112. The four cam intervals

The theoretical opening point of the valve is at the end of the opening ramp since this, in theory at least, is the time when the valve clearance has been taken up. In the first interval the valve is accelerated to maximum velocity, in the second interval the valve is decelerated to zero velocity at the top of the lift. Maximum negative acceleration occurs at this point and the force to provide it is supplied by the spring. The third and fourth intervals are respectively an acceleration to maximum closing velocity and a deceleration to zero velocity at the start of the closing ramp. The force to provide this final deceleration is of course provided by the cam.

If the valve train (the tappet, push-rod and rocker, in the majority of engines) were perfectly rigid and incompressible, half the problems of valve design would disappear. The remaining problems would be associated with the valve springs and the vibrations produced in the coils by their own inertia. Most of the springiness of the valve train comes from the push-rod. How bad it can really be is shown in Fig. 113. This is a problem that can be partially

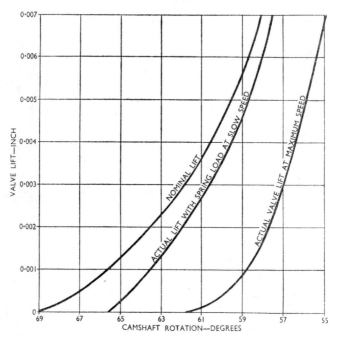

Fig. 113.   Variations in opening conditions due to spring load and inertia
Centre line of cam nose is at o camshaft degrees
Clearance 0·015 inch at valve 115 deg. period multi-sine wave cam
Nominal timing—0·022 inch clearance at valve

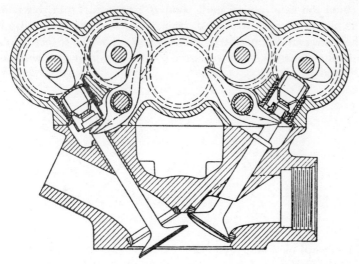

Fig. 114.   Experimental Norton motor-cycle desmodromic cylinder head

solved by the use of overhead camshafts and completely solved by
the provision of a *reliable* form of desmodromic valve gear (see Fig.
114).

The actual valve motion

The use of the stroboscope and the high speed camera has shown
that the motion of a valve at high engine speeds is seldom the same
as the theoretical motion defined by the cam profile. It is precisely
this failure to make the valve follow the designed motion that has
led designers to adopt the modern overhead camshaft design
using piston-type tappets. Others, perhaps a little bolder than the
rest, have attempted the more difficult alternative of the desmodromic gear (from the Greek roots meaning 'chained to a track').
The track is not difficult to provide, but the problems of differential
expansion and the maintenance of fine running clearances have
prevented the general adoption of this valve gear for production
cars.

Flexibility in the valve train is the enemy of precise valve operation. Fig. 115 shows a typical trace of the valve motion on a push-
rod o.h.v. engine at 5000 r.p.m. The valve starts to lift at *A*, but,
because the push-rod and rocker arm bend, the valve motion lags

behind the cam motion until point *B* is passed. From this point the inertia of the valve and its train, plus the energy stored up in the spring of the push-rod and the rocker, begin to act in the opposite direction, since the cam is now beginning to decelerate

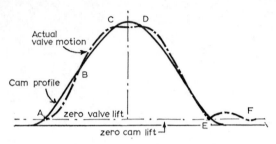

FIG. 115.   Trace of actual valve motion

the upward motion. At point *B*, then, the tappet loses contact with the cam face and does not regain it until point *C* is reached. A wear zone just ahead of the cam nose is an indication of this effect. The cam lands heavily on the elastic valve train at point *C* only to send it back again into space like a diver on a springboard. The cam catches up again with the errant valve train at *D* and in the particular trace shown it is seen to bounce again at *E* to the detriment of any clear definition of the point of valve closing.

Spring surge

Valve spring surge is caused by the inertia of the spring coils and the tendency of all springs to have a natural frequency of vibration. The acceleration imparted to the coils during the opening and closing of the valve causes waves to travel backwards and forwards along the length of the spring. When the frequency of these forced vibrations coincides with certain harmonics of the natural vibration frequency of the spring the ripples travelling along the length of the spring increase in magnitude, so much in fact that the ripples or surges can be seen with the naked eye when a stroboscope is used. At its worst, valve spring surge becomes valve bounce, i.e. the end of the valve stem loses contact with the valve rocker. In a mild form it leads to an increase in the stress range to which the valve spring is subjected. In more severe forms the fatigue life of the spring is drastically reduced. Breakage of the bottom coil is an indication of this condition. The use of double and triple valve springs is a useful safety measure against a dropped valve, but the individual springs are still free to surge inside each other. However, since the harmonics of the different springs do not coincide, the tendency for the valve to 'chatter' on the seat is much reduced.

Higher engine speeds and/or higher valve lift call for stronger valve springs to counteract the increased inertia of the moving parts, but the tuner must resist the temptation to step up the spring strength indiscriminately. Every pound of additional spring load is a pound increase in the load on the cam face.

Cam-face wear

Scuffing of the case-hardened cam-face and of the chilled cast-iron tappet face is a problem that arises with high-speed engines and an increase in valve lift should not be undertaken lightly. When the original case-hardened skin of the maker's standard cam has been completely ground away in the profiling of the new high-lift cam (as shown at the left of Fig. 116) it is essential to re-treat

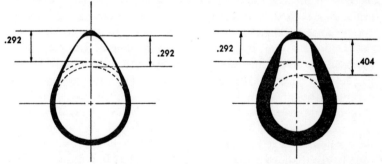

Fig. 116. Most cams are reground from the stock contour. The black portion shown is ground off. The cam on the left has the same lift as the stock cam but a different contour. The '404' cam on the right has been ground to give a higher lift as well as a faster rate of lift than the stock cam.

the camshaft after the re-profiling operation. The responsible specialists in the cam-grinding business have developed excellent hardening techniques for their re-ground camshafts. In the past the writer has seen special re-ground camshafts in which no re-heat treatment was carried out after the re-grind. Many of these older re-grinds were simply a reduction in diameter of the base circle with a blending-in of the new base circle and the old flanks of the cam lobe. For a side-valve Austin Seven such a crude treatment would suffice, since the cam-face loads were so low. On a modern high-speed sports car engine unhardened cams would not last very long.

The majority of the sports car re-grinds supplied by Iskenderian are treated in one of two ways. The steel shafts are carburised. The cast-iron ones are induction hardened and finally Parkerised after grinding. For certain competition camshafts Iskenderian has developed the hard-face overlay technique. In this process a hard alloy is deposited on the cam-shaft, then the desired profile is obtained by grinding. Superior resistance to scuffing is claimed for this process. It has been used under the most severe conditions of extremely high lift and high r.p.m. on many successful drag machines.

When to fit a high-lift camshaft

It is just as important to know when *not* to make a change from standard. Sometimes the fitting of a camshaft with a moderate increase in lift and duration will completely transform a car from a plodding hack to a thrilling racehorse. Sometimes the change in performance is difficult to find without careful stop-watch work and the customer is soon beginning to wonder how he can escape the financial burden of his folly. Here then is a rough guide for the professional tuner. Look for a worthwhile improvement if the existing ratio of valve lift to throat diameter is less than 0·24. Where the present ratio is as high as 0·28 some useful gain in top-end performance can still be made by fitting a higher lift camshaft, but the economics of it become marginal. Negligible improvement is usually given above a ratio of 0·32 and the loss in bottom-end per-formance, especially when a 'real wild' valve timing is used, will sometimes completely ruin a car for street use.

In case of doubt the advice of the factory competition depart-ment should always be sought. The man who really wants to win a race and will take a sporting chance on rapid cam wear, even on dropped valves, will at least be doing it with his eyes open after he has discussed the reasonable limits with the people who make the engine.

In Appendix 1 at the back of the book will be found a list of camshafts that should serve as a guide to the factory options and specialist grinds available for the more popular sports cars. In comparing camshafts it must be remembered that a direct comparison is not always possible from the lift and timing data. Some cam profiles give extremely rapid lifts (such as the famous 'Isky 404' shown in Fig. 116) and could give a greater valve area/time integral than an alternative cam profile of much greater duration, but with a less drastic opening and closing acceleration. Another factor that can be misleading is that some camshafts are checked for timing with a greater clearance than the normal running clearance. The quoted valve timings are therefore slightly later at the opening and slightly earlier at the closing than the actual valve timings with the correct running clearances. The valve timing on Porsche cams, for example, is checked with a valve clearance of 0·040 inches.

BLOWERS

A four-stroke engine devotes two of its strokes to the duties of a compressor. Mixture is drawn in on the induction stroke and compressed on the compression stroke. The power we can extract from a given size of engine (heat load and bearing limitations apart) depends almost entirely on the mass of air that the engine can compress into, and reject from, the working cylinder per unit

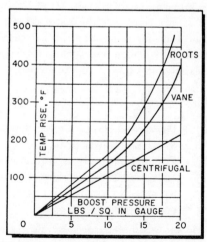

FIG. 117. Approximate intake air temperature rise produced by the three types of superchargers at various boost pressures.

of time. When we supercharge an engine we are making the engine act as the second stage of a two-stage compressor. The first-stage compressor, blower, or supercharger feeds air or mixture to the engine at a pressure substantially in excess of atmospheric pressure. This supercharge pressure might be $1\frac{1}{3}$ to $1\frac{2}{3}$ atmospheres absolute for a sports car and as high as 4 atmospheres absolute in the case of some racing cars in the past. The density increase will not be quite in proportion, since the compression of the air in the super-charger results in a substantial rise in temperature (see Fig. 117). There is also a certain allowance to be made for the power ex-pended in driving the supercharger.

In the range we are considering as applicable to the majority of sports cars, i.e. from 5-10 lb. per sq. in. of boost measured at the induction manifold, the nett increase in power is almost in direct proportion to the increase in induction pressure. Thus a boost pressure of 6 lb. per sq. in. at 6000 r.p.m. will give an increase in power of $\dfrac{6}{14\cdot7}$ or 41 per cent. With a naturally-aspirated engine the induction pressure at maximum power r.p.m. is not, of course, 14·7 lb. per sq. in., but a much lower pressure, around 12 lb. per sq. in. Thus the boost pressure of 6 lb. per sq. in. at 6000 r.p.m. represents an absolute pressure of 20·7 lb. per sq. in. in the manifold, or an increase of 72 per cent over the naturally aspirated engine.

The writer's rule then for estimating the increased power to be given by a mechanically-driven supercharger (as distinct from a turbocharger or exhaust gas driven supercharger) is as follows:

$$P_s = \frac{P_n \times p_b + 14\cdot7}{14\cdot7} \tag{14}$$

where P_s = Power with supercharger, measured at r.p.m. of original power peak.

P_n = Original naturally aspirated maximum power.

p_b = Induction pressure given by supercharger at r.p.m. of original power peak, lb. per sq. in. gauge pressure.

The use of a moderate supercharge may also extend the power curve by about 10 per cent, but the safe mechanical limitations of the engine dictate how much advantage we dare take of this increase in operating speed range.

The mechanics of supercharging

Mechanically-driven superchargers are of three types, the Rotary Vane, the Roots and the Centrifugal.

Vane type

The working of this type is shown by Fig. 118. The sliding vanes are rotated by a cylindrical rotor which runs in end bearings that are set eccentrically to the centre-line of the cylindrical housing. Rotation of the rotor by means of any suitable drive from the engine (usually a belt drive from a pulley on the crankshaft) causes mixture to be trapped between vanes as they pass the intake port and this mixture to be compressed gradually as the vanes rotate to the discharge side. In the American Judson supercharger, which is

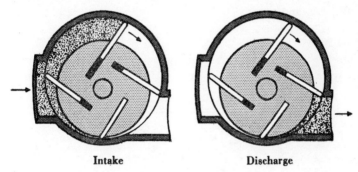

Intake   Discharge

FIG. 118. Schematic drawing of Judson vane type supercharger

shown schematically in Fig. 118, the vanes are given a decided off-set relative to the rotor centre-line. This permits the use of longer slots in the rotor and thus contributes to the long life of the vanes. The vanes are of phenol-formaldehyde resin impregnated linen, a material similar to our British 'Tufnol'. This material is tough, hard-wearing and silent in operation. Two relief slots milled in the pressure side of each vane vent the air trapped below the vanes and minimise the power loss from this source. The blade tips are rounded and are claimed to run on a protective air film at high speed. Lubrication is still necessary and is provided by a special Judson oiler that draws its supply from a half-gallon tank.

The vane-type supercharger has been developed to a high stage of reliability and when installed in a sound engineering manner will give excellent service for many years. Three excellent designs can be recommended.

Arnott

Carburettors Ltd, Grange Road, London N.W.10, England.

Judson

Judson Research and Manufacturing Co, Conshohocken, Pa., U.S.A.
and

Performance Equipment Company Ltd, Sandford Street, Birkenhead, England.

Shorrock

Shorrock Superchargers Ltd, Church Street, Wednesbury, Staffs., England.

Roots type

The Roots blower is really a gearwheel pump, with two, three or four teeth or lobes to each wheel. In the more common two-lobe

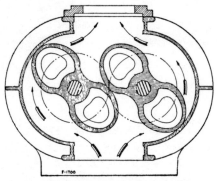

FIG. 119. Principle of Roots supercharger. Two lobes are geared together and rotate in exact phase with very little clearance, without actually touching each other.

type as shown in Fig. 119 the two rotors are interconnected by two meshing spur gears that run in a separate compartment outside the air compression section of the supercharger. The rotors are given a very small working clearance, but this must be sufficient to ensure that no contact can occur between the rotors or between a rotor and the outer casing at all times. Lubrication is required for the gears, but none is required inside the rotor chamber. It will be seen that compression does not occur until the volume of mixture trapped between the rotor and the outer casing is released to the mixture at higher pressure in the delivery port. Compression then occurs under the action of a back-flow of mixture into this space between the rotor and the outer casing. Since this back-flow occurs at the moment when the leading edge of the rotor clears the edge of the delivery port, the compression is rather explosive in character. This, plus the whine of the gears, tends to make the Roots blower noisier than the vane type.

Careful design of the induction system is also important when a Roots blower is used, since the pressure waves induced by the successive delivery pulses can impair or enhance the blower performance at certain critical speeds, in the manner of a tuned induction system. Proprietary kits are of course bench tested, but the amateur tuner who is making up his own installation must not overlook the possibility of trouble from this source.

Suppliers of three well-known makes of Roots blower are listed below.

Borg-Warner

M-D Blowers, Inc, 100 Fourth Street, Racine, Wisconsin, U.S.A.

Marshall-Nordec

North Downs Engineering Co., Westway, Caterham, Surrey, England.

and

Speedsport, 4868 Milwaukee Avenue, Chicago 30, Ill., U.S.A.

Wade

Wade Engineering Ltd, Gatwick Airport, Horley, Surrey, England.

Centrifugal type

The centrifugal supercharger is well known in the field of aero-engines, but apart from the unrewarding attempt to match its characteristics to the demands of the Mark I B.R.M., this type has seen little use on European cars. The reason is soon apparent when one considers that the pressure supplied by a centrifugal blower *varies as the square of the speed.* Thus if we design to provide a boost of 6 lb. per sq. in. at 5000 engine r.p.m. we will only get a boost of 1½ lb. per sq. in. at 2500 engine r.p.m. The American McCulloch blower overcomes this difficulty by means of an ingenious variable speed drive. By automatic variation in the ratio between engine speed and supercharger speed the boost pressure rises rapidly to

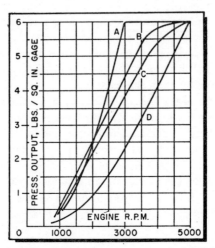

Fig. 120.   Pressure output curves: A. McCulloch, variable ratio; B. Vane-type, fixed
ratio; C. Roots-type, fixed ratio; D. Centrifugal, fixed ratio.

the limiting value. From the middle of the speed range, say 3000 engine r.p.m., the blower speed stays constant, whatever the engine speed. This gives a constant boost at all pressures above 3000 r.p.m. This effect is shown in Fig. 120, where the boost curves of typical examples of the three types of blower are compared.

McCulloch Supercharger

Paxton Products, 827 W. Olive Street, Inglewood, California, U.S.A.

SUPERCHARGER TUNING TIPS

Heat rejection

When we extract more power from an engine by increasing the compression ratio we burn the same amount of fuel as before, but we burn it more efficiently. Consequently we reject less heat to the water jackets and less heat to the exhaust.

When we supercharge an engine it is a different story. If we make no change in the compression ratio and at the same time increase the air consumption by 50 per cent, we increase the fuel consumption by 50 per cent and reject about 50 per cent more heat to the water jackets and about 50 per cent more heat to the exhaust gases. If we are forced to lower the compression ratio to avoid detonation, the heat rejection rate is increased still more. The moral of this story is simple. Look to the cooling system first. If it is known that the radiator capacity was only just adequate with the original unsupercharged engine it may become necessary to fit a deeper radiator block after supercharging. If any doubt exists about the quality of the exhaust valve steel, then a change to a higher grade material should be made. Exhaust valve steels are discussed in Chapter Thirteen and a short list of inlet and exhaust valve materials is also given.

Blowing through the carburettor

It is universal practice to-day to fit the carburettor on the atmospheric side of the supercharger, but several racing cars in the past supercharged through the carburettor, which necessitated a sealed and pressure-balanced float-bowl. Mercedes Benz used this method for many years and their engineers stated that this was the only logical position for the carburettor. The argument in favour was that this layout would give a quicker response to the opening of the accelerator immediately following a period of closed throttle operation, since the blower would build up pressure on the up-stream side of the closed throttle-plate. When the accelerator was opened again this build-up of pressure would help overcome any flat-spot. Occasionally one meets an installation where a slight hiatus occurs, but in general the problem is one of carburation. This older Mercedes Benz method involves certain added complication and the writer does not think the results justify the trouble.

Carburation

Do not neglect to step up the carburettor venturi size, even when the degree of supercharge is small. An increase in venturi area approximately equal to the increase in absolute manifold pressure should be provided. A blower is a wonderful mixer for air and fuel. It gives mechanical agitation and supplies heat to vaporise the fuel. With such a device following the carburettor we can afford to neglect good atomisation at the venturi and concentrate on reducing the pressure drop through the carburettor.

Whenever possible it pays to take advantage of an available supercharger kit. Apart from the problems of finding a suitable drive, and adequate space for it, the tuning of the carburettor can in itself become a major problem, especially if the length and capacity of the manifold is of such dimensions as to cause blower surge at certain speeds. Not enough is known about this problem, but the tuner can at least be warned of its existence. American hot-rodders have been fitting superchargers to slingshots and all manner of dragsters for many years, but their problems are simplified by the very nature of their sport. A sports car demands a wide range of tractability from its power unit and is therefore more exacting in its carburation demands.

The drive

There is no need to look further than the simple V-belt drive when installing a supercharger. Roller-chains and gear trains can of course be used, but a belt drive is reliable, efficient and makes the least noise. Here are four simple rules. Don't use a belt speed above 8000 ft. per minute. Don't use a pulley diameter below 5 inches. Don't use a contact arc of less than 120 degrees on any pulley. Failure to comply with any of these three simple rules will lead to short belt life. Finally, don't use cast-iron pulleys. The pieces can make a very unsightly hole in the bonnet.

Spring-loaded idler pulleys are sometimes used but the writer prefers to use an idler that can be adjusted in a slide to take up tension and then locked in position. The reason for this is as follows: Sometimes during deceleration the supercharger decelerates at a greater rate than the engine. In this condition the supercharger is driving the engine and the side of the belt on which the jockey pulley is fitted becomes the drive side. With a spring-loaded idler pulley the sudden application of drive to this side causes the idler to jump backwards and forwards and the belt to snatch. This reduces belt life. Multiple belts are usually necessary to transmit the horse power involved in driving a blower. Figures for the powers absorbed by the particular blower can always be obtained from the makers. Here are some belt capacities for driving blowers. A $\frac{1}{2}$ inch wide belt will transmit 5 b.h.p. at 6000 r.p.m.; a $\frac{21}{32}$ inch wide belt will transmit 7 b.h.p. at the same speed and a $\frac{7}{8}$ inch wide belt will transmit 12 b.h.p. Thus if our blower absorbs 20 b.h.p. at 6000 r.p.m. we could use either four $\frac{1}{2}$ inch, three $\frac{21}{32}$ inch or two $\frac{7}{8}$ inch belts.

A recent development in belt drives is the internally toothed 'timing belt'. This belt is of rubber with flat bars or teeth on the inner face. These teeth engage corresponding teeth on the flat pulleys. The outer section of the rubber belt is reinforced with steel wire. At 8000 ft. per minute belt speed one of these belts, only 1 inch wide, will transmit 20 b.h.p. It is claimed by the makers that these belts will operate satisfactorily at belt speeds up to 15,000 ft. per minute.

Valves and ports

Because a blower has been fitted is no excuse for throwing away part of the boost pressure in overcoming the drag of rough and irregularly shaped ports or undersized valves. Work on the head to improve air-flow is just as important as on an unsupercharged engine. The exhaust side in particular should be opened up as much as possible. With a boost of 6 lb. per sq. in. we will be inspiring about 40 per cent more air into the engine and passing a proportionate increase in gas flow out of the exhaust system. Where possible pipe sizes should be increased to prevent undue build-up of exhaust back-pressure.

Camshafts

It is not essential to make any changes from the standard valve timing when modest boosts are used. When a change in camshaft is made it is customary to reduce valve overlap to reduce fuel losses down the exhaust pipe.

Compression ratio

A boost pressure of 6 lb. per sq. in. will raise maximum cylinder pressures and bearing loads by the equivalent of about 2 compression ratios. It will give much more power than a compression ratio increase of 2 numbers, so why be greedy? For the sake of reliability a reasonable limit for the typical engine would be a compression ratio of 8 to 1 when using a maximum boost of 6 lb. per sq. in.

Ignition system

A high voltage coil will usually be adequate for speeds up to 7000 r.p.m. and boosts of 6 lb. per sq. in. on four-cylinder engines, but on six-cylinder engines a change to magneto ignition or the fitting of a transistorised coil system might become essential to realise the full potential of the boosted engine.

As a general rule one can count on the optimum ignition timing being a little later with a supercharged engine. Since the temperature and density of the mixture is higher, flame speeds increase and the use of the standard ignition timing would permit a harsh pressure rise to occur before T.D.C. About 5 degrees retard from normal at full power would be typical for an engine boosted to 6 lb. per sq. in., with a normal timing at low boost and speeds below 2500 r.p.m.

Sparking plugs

Plugs can be a problem on a supercharged engine. The use of supercharge, even to a moderate degree, calls for a change to a much cooler plug, i.e. one of higher heat capacity. Such a plug will tend to short out in city traffic driving if the driver is not constantly alert to the problem. Some drivers will disagree with this, especially if they are unconscious of their youthful exuberance. Nevertheless an itchy accelerator foot is the best cure of plug fouling on supercharged engines.

To help maintain a good spark at high r.p.m. the plug gap should be reduced to 0·18 inch and the mixture enriched about one ratio.

TURBOCHARGING

The use of the energy in the exhaust gas of an internal combustion engine to drive a supercharger is no novelty. Perhaps the earliest application was the conversion of a Liberty aero-engine to the system by Jesse G. Vincent in 1928. The latest production application to cars, developed between 1956 and 1961 by the General Motors Corporation in America stems from the earlier work by the Garrett Corporation who have been making automotive and locomotive turbochargers since 1946.

The basic principle is shown in Fig. 121. The exhaust gases pass
by way of the manifold to the turbine wheel. Gas flow is tangential
at first from a scroll chamber, which may or may not have guide
vanes. The gases strike the vanes of the turbine wheel and then
flow inwards to the exit pipe. The expansion of the hot gases from
the relatively high pressure of the scroll chamber to the near-
atmospheric pressure of the exit pipe provides the power to rotate

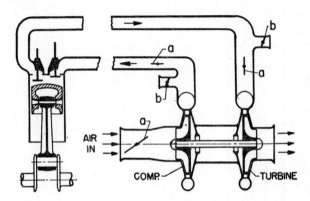

FIG. 121. Turbocharger installation showing two ways of con-
trolling the output. (*a*) is by throttling; (*b*) is by by-passing.

the centrifugal compressor which is mounted at the opposite end
of the turbine shaft. The design of the compressor is orthodox,
with radial vanes and an outward flow.

Control of the system can be by throttling, by by-pass of exhaust
gas, by by-pass of compressor delivery, or by a combination of
both. At this time the G.M. turbocharger is available on two cars,
the Corvair Monza Spyder and the Oldsmobile Jetfire. The
systems differ chiefly in the method of control. The Corvair
turbine wheel has 11 blades and is 2·97 inches in outside diameter.
The designed maximum speed is 70,000 r.p.m. On the Oldsmobile
installation the diameter has been reduced to 2·4 inches and the
normal maximum speed raised to an awesome 90,000 r.p.m. The
turbine wheel is a high nickel cobalt steel forging with adequate
hot-strength for the operating temperature at full-load of approximately 1500°F. On the Corvair the compressor has a 14 bladed
impellor of 3·0 inches diameter, cast in aluminium. The Olds

compressor is smaller, to match its smaller turbine, being only 2·5 inches in diameter.

The Corvair is the simpler installation. Boost only begins to build up to any measurable pressure when an engine speed of 2000 r.p.m. is exceeded (see Fig. 122) and from this speed upwards the boost rises rapidly to a value of 10 lb. per sq. in. at 4000 r.p.m.

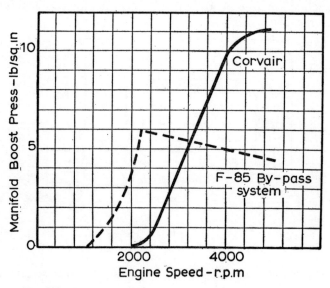

Fɪɢ. 122.   Boost characteristics on Corvair and Oldsmobile turbocharged engines

No special controls are provided on the Corvair design, other than the natural limitations of the system. At engine speeds above 4000 r.p.m. the exhaust back-pressure must rise to a very high figure (the actual value has not been disclosed by the G.M. engineers). The back-pressure rises so high that it acts as a control on the breathing of the cylinders and this in itself acts as a built-in overspeed safety device. The Olds installation with a smaller turbine wheel to drive builds up to a back-pressure of 13 lb. per sq. in. at 4800 engine r.p.m. The boost at this speed is controlled to 4-5 lb. per sq. in. (see Fig. 122). The Olds turbocharger is designed to give full boost at just over 2000 engine r.p.m. At higher mass-flows through the turbine a by-pass valve opens and passes the surplus gas around the turbine to the exit side. At peak power at 4600 r.p.m. engine speed approximately one-half of the exhaust gas is by-passed. The by-pass valve is controlled by a large diameter diaphragm unit acted on by boost pressure. Failure of the by-pass valve to open is safeguarded by a second diaphragm-controlled butterfly throttle in the induction system.

When fitted with the turbocharger the Corvair engine is delivered with a reduced compression ratio. The Oldsmobile Turbo-Rocket engine has a compression ratio of 10·25 to 1 and a special alcohol-water injection system is incorporated into the Rochester side-draft carburettor. This system meters increasing amounts of the special anti-knock fluid as the boost pressure rises.

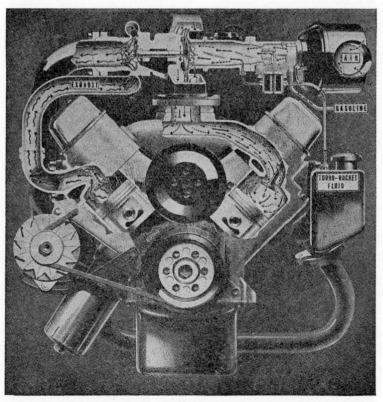

Fig. 123.   Exhaust turbocharger on Oldsmobile F-85

When used with a sports car engine the turbocharger has one serious drawback. When a driver of any road vehicle puts his accelerator to the floor he expects an instant response; on a sports car he demands it. Unfortunately the boost pressure on a turbocharger takes time to build up. A road-test of the Corvair Monza Spyder showed a lag of about 4-5 seconds for the boost to catch up with its true steady reading at the particular engine speed. For example, if the throttle is held at the position that gives a steady 4000 engine r.p.m. in third gear (approximately 56 m.p.h.), then the throttle is suddenly opened wide, it takes about 4 seconds for the full-throttle boost of 9-10 lb. per sq. in. to be shown on the boost gauge on the instrument panel. In the part-throttle condition the turbine/compressor unit will be rotating at about 20,000 r.p.m. Full-throttle boost at 4000 r.p.m. requires a turbine speed of about 55,000. Even with such a small diameter turbine wheel and compressor impellor one could hardly expect build-up of about 35,000 r.p.m. in much less time. The overall effect of this delay is to make this Corvair hardly any more accelerative than the unblown version.

The Olds system goes a long way towards overcoming this disadvantage. By developing a moderate boost at 2000 r.p.m. engine speed (where the Corvair boost is zero), and by a reduction in the rotating mass of the turbine wheel and the impellor, the lag time is reduced to about one second.

CHAPTER THIRTEEN

The Mechanics of Modification

IN EARLIER CHAPTERS we have considered many aspects of the technique of engine modification. In this final chapter we shall discuss a few of the practical problems one meets when applying these techniques. This chapter is something of a hotch-potch with very little logic in its arrangement, but a liberal use of headings will help the reader to find a particular subject quickly. It is assumed throughout this chapter that the reader, whether he be owner or professional mechanic, is tuning an engine for competition work. A modern rally, especially in the modified class, comes inside this definition.

CYLINDER HEAD WORK

Compression ratios and how to change them

Whether we remove metal from the cylinder head face or fit high compression pistons it is still necessary to measure the compression ratio of the engine, both before and after modification. We also need to know how much metal to remove from the head or how much to increase the compression height on the new set of pistons to obtain a desired increase in compression ratio.

$$\text{The compression ratio } R = \frac{V+v}{v} \qquad (15)$$

$$\text{where } V = \text{the swept volume of one cylinder,}$$
i.e. the engine capacity in c.c. divided by the number of cylinders.
$$v = \text{the clearance volume in c.c.}$$

For our particular purpose, the equation is more convenient rearranged in the following form:

275

$$v = \frac{V}{R-1} \qquad (16)$$

Thus, if we wish to raise the compression ratio from R_1 to R_2 the clearance volume v_1 must be reduced to v_2.

This reduction in clearance volume is given by:

$$v_1 - v_2 = V\left(\frac{1}{R_1-1} - \frac{1}{R_2-1}\right) \qquad (17)$$

For example, if the swept volume of one cylinder,

$$V = 400 \text{ c.c.}$$
$$R_1 = 8\cdot0 \text{ to } 1$$
$$R_2 = 9\cdot5 \text{ to } 1$$

The required reduction in clearance volume,

$$v_1 - v_2 = 400\left(\frac{1}{7\cdot0} - \frac{1}{8\cdot5}\right)$$
$$= 10\cdot08 \text{ c.c.}$$

Measuring the clearance volume

The volume of the combustion chamber can be measured directly by filling the volume with thin oil. Some prefer to place the cylinder head face down on a sheet of plate glass and to fill the combustion chamber through the plug hole from a graduated measuring cylinder. The cylinder head face should be lightly covered with Vaseline to provide a good seal between it and the glass. The filled level of the oil in the plug hole should be slightly higher than the plug reach to allow as close as one can judge by eye for the interior volume of the plug. The writer prefers to place the cylinder head upside-down on the work-bench with the plug tightened in position and to fill the head from the open side of the combustion chamber. Before filling the cylinder head the face must be checked for perfect level by means of a spirit level, using packings under the head to get the desired level. The combustion chamber space of the cylinder to be checked is then filled with oil, a steel rule being placed across the head face to verify that the oil level is exactly flush with the head face. A glass measuring cylinder graduated in c.c. is used to obtain the exact volume measurement taken to fill the combustion chamber space.

In all cases where a flat-topped piston is used and the piston at T.D.C. is level with the cylinder face, an additional volume, representing the volume occupied by the head gasket in its fully tightened condition must be added to the measured volume to find the true clearance volume. When the piston crown is domed, or in the shape of a truncated cone, this volume must be calculated from its geometric considerations and the volume of this intrusion into the combustion chamber deducted from the measured volume. When the piston crown is of indeterminate shape one has no alternative but to measure the clearance volume with the head fitted to the block. Before the head is fitted, however, the piston in the cylinder to be measured should be carefully set to T.D.C. and the top land sealed with Vaseline or thick grease, care being taken to keep the quantity used to a bare minimum. As the measured quantity of oil is poured into the plug hole, a wire stirrer should be twisted round inside the combustion chamber from time to time to prevent the accumulation of bubbles in corners.

How much metal to remove

Having measured the clearance volume of the unmodified cylinder, we now know V, the swept volume of one cylinder and v, the unmodified clearance volume of this same cylinder. From equation (15) we can therefore calculate R_1, the original compression ratio. Having chosen the value of the modified compression ratio, R_2, we can use equation (17) to calculate $v_1 - v_2$. This gives us the new clearance volume, v_2.

Now, using the method preferred by the writer, in which the cylinder head is placed upside-down and filled with oil from the open end of the combustion chamber, we can now pour the exact quantity, v_2, into the combustion chamber, then measure with a depth-gauge how far below the machined surface of the head the new level of oil is lying. This measured distance is the amount of metal to be machined from the head to give the new compression ratio.

High compression pistons

In general one should never remove more than $0.125''$ from the head face of an engine. For engines of smaller capacity than 1500 c.c. one cannot always assume that this amount can be removed with complete safety. In all doubtful cases the advice of the competition department can be sought. When a greater change in ratio is required high compression pistons can be fitted. When these are not available from the engine maker the next people to be tried are the larger replacement piston manufacturers such as Hepworth and Grandage of Bradford, Yorkshire, England. When custom-built pistons are the only answer the Martlet Company of Brooklands, Surrey, or Jahn's Quality Pistons Inc, Los Angeles, are ready and willing to meet this specialist need.

Occasionally, when a very high compression ratio is required, the addition of extra volume to the piston crown can result in a

FIG. 124.   An assymetric piston crown.
Beware distortion of ring-belt!

combustion chamber shape that is nearly all squish area, an undesirable feature in a high compression engine. A large lump of metal on the piston crown is again not at all desirable from the heat conduction standpoint (see Fig. 124). If the mass of metal is extremely assymetric, distortion of the piston crown is quite possible. This could interfere with the sealing of the top ring, with possibly disastrous consequences. It is sometimes necessary to make a compromise to avoid this danger. Some of the decrease in chamber volume can be provided by the special high-domed piston, while the rest is obtained by machining metal from the cylinder head face.

Machining methods

The removal of the chosen amount of metal from the head face should only be entrusted to a competent machine shop. Surface grinding will give the best surface finish, but an acceptable finish can be given by milling. Beware of the small automobile shop that offers to carry out the work on their head re-facer. With these machines, intended only for the quick re-facing of heat-warped heads, the head is held by the machine operator and pushed across the grinding table. Any attempt to remove more than 0·005″ of metal with such a grinder is liable to ruin the head.

Effect on the valve gear

The removal of metal from the cylinder head can interfere with the operation of other components. On an o.h.c. cylinder head with gear drive to the camshaft one can seldom attempt to raise the compression ratio by head machining, since the gear centres cannot be changed. Even when a chain drive is used, many makers do not recommend the method. Every case must, of course, be examined on its merits. For the popular push-rod operated o.h.v. engine there is usually no objection to machining the head face in moderation. All that is usually necessary to compensate for the reduction in head thickness after the head has been machined is the provision of steel packings placed under each rocker shaft pillar to maintain the rocker shaft at the original centre-line distance from the camshaft. An alternative is to shorten the push-rod tubes by an amount equal to the depth of metal machined from the head face. When cap nuts are used on the cylinder head, flat steel washers should be placed under each nut to prevent the cap nuts from bottoming.

Matching the head volumes

All re-shaping and polishing of the combustion chamber will have been carried out before the final face machining to increase the compression ratio. The perfectionist (and all the well-known names in the tuning world could not have survived without this quality) always finishes the work on his cylinder head by matching the clearance volumes. All clearance volumes are measured carefully and, when the combustion chamber with the largest volume is known, metal is ground from the others until all the clearance volumes are matched. This can be tedious work! Some of the professional tune shops save time in the matching operation by the use of sheet-metal templates. In this way they check the depth of different parts of the combustion chamber below the joint face. This cuts out the time-consuming liquid measuring operation, but the work involved in making up templates is only warranted when at least half a dozen heads of the same type are to be worked.

Maintenance of head clearances

On some high output engines very little clearance exists between the valve heads and the piston crowns. Recesses are sometimes machined in the piston crowns to accommodate the valve head at T.D.C. It is essential when major machining work has been carried out or a new camshaft has been fitted with increased overlap that a careful check be made to see that adequate clearances have been maintained. A simple method is to place strips of modelling clay across the piston crowns on all pistons. The head is then fitted with its correct gasket and tightened down, tightening the nuts in the correct order and to the recommended torque. The engine is then turned over slowly by hand for two complete revolutions. When the head is removed the minimum clearance of each piston crown can be seen where the modelling clay is most flattened. This clearance should never fall below 0·040″.

Work on the ports

The theory of porting has already been well covered. Here we are interested in the practical aspects of the work.

Where larger valves are to be accommodated a milling cutter must be used with a pilot centralised by the valve guide. Sometimes, as in the bath-tub head, a side cutter is necessary to increase the combustion chamber clearance around the new larger valve head. When a milling machine is not available resourceful tuners have been known to open up the valve throats with a portable grinder, but the method is not normally to be recommended.

At the completion of the milling or boring operation to enlarge the valve throat, we are usually left with a step on one side of the port wall where the enlarged bore stops. This step must be ground away to leave a smooth transition between the enlarged throat and the old port. Unless our calculations indicate that the ports are already large enough to maintain gas velocities that are well matched to the valve throat size, our next task is to open up the ports. Normally these are maintained of equal cross-section from throat to manifold face. Sometimes when the wall thickness is known to be large we can risk grinding out a tapered port, with a gradual enlargement from throat to manifold face. There is never any risk in grinding away $\frac{1}{32}''$ of metal from all sides. Sometimes as much as $\frac{1}{16}''$ can be ground away, giving an increase on a circular port of $\frac{1}{8}''$ in diameter. This is an increase of 21 per cent in area on an original port diameter of $1\frac{1}{4}''$. The removal of any more wall thickness than $\frac{1}{16}''$ involves a slight risk of striking a thin porous section of the casting. The risk is not great but occasionally a core has slipped in the casting operation and the port walls are thicker than nominal on one side and thinner than nominal on the other. In the hundreds of cylinder heads that the writer and his staff have modified only two have been scrapped by porting work.

Excellent grinding heads complete with flexible drive shafts suitable for porting work are made in England by Wolf Electric Tools and Black & Decker. In America the makers are too numerous to list. The free speed of the grinding head should be at least 10,000 r.p.m. Rotary files and grinding stones are available in a wide range of sizes, shapes and roughnesses. For final polishing one can use emery cloth mops, followed by cloth mops impregnated with a fine carborundum polishing compound. The degree of polish is less important than the extent of surface irregularity. A highly polished mirror-like surface that is full of ripples or variations in cross-section would offer a greater resistance to gas flow than a matt surface free from any irregularities.

INCREASING THE LIFE OF COMPONENTS

Crack detection

The detection of cracks in the surface of a highly stressed engine component is, paradoxically enough, an excellent way of increasing the life of an engine, even of its individual components. A small crack in a crankshaft or connecting rod, discovered in time, will lead to the replacement of the particular component. In this way the life of many other expensive components, such as the cylinder block and a full set of rods and pistons has been spared. The breaking of a con rod at full speed has sometimes called for a new bonnet and wing valence!

The Magnaflux Corporation of America were pioneers in the commercial application of magnetic detection of cracks. A magnetic field must first be set up in the part to be inspected. The part is then treated with a solvent containing finely divided iron particles. The presence of a crack interrupts the magnetic field producing a localised intensification of the lines of force. This attracts a higher concentration of iron particles around the edges of the crack and reveals its presence to the naked eye. A more searching test still is the Magnaglow technique developed by Magnaflux. In this method a fluorescent paste, dispersed in oil, serves to make the magnetic particles fluoresce when the part under examination is placed under a source of 'black light'. The method is very effective, but only of course on cast iron and mild steel and many of their alloys. Austenitic steels, aluminium and copper alloys, being non-magnetic, cannot be inspected in this way.

Dye penetrants are usually used to inspect non-magnetic components. One good method we can recommend is to brush the surface of the part with a mixture of one-third paraffin (kerosene) and two-thirds of S.A.E. 30 grade oil. The component is then wiped clean and immediately painted with a zinc oxide paint with a methanol solvent. The presence of a crack is revealed by a discolouration of the white surface. A handy method available in America is now being marketted by the Magnaflux Corporation. This involves the use of three spray cans. The first is a cleaner, the second is a dark red penetrant. After spraying with the penetrant the part is wiped clean. Application of the final fluid, the developer, brings up the presence of a crack as a vivid red line. No special lighting is required for this method. It can therefore be used on parts in situ, on suspension links, steering arms and idlers and, in fact on any component with a smooth finish.

Shot-peening

Machine shops that cater for the motoring enthusiast are usually equipped with shot-peening equipment. The surfaces of steel components that are normally subjected to high alternating stresses are subjected to a high velocity blast of small steel pellets. This modifies the crystal structure at the surface of the metal to form a tough skin that has an improved resistance to the formation of surface cracks.

Balancing

The term balancing, when applied to a reciprocating engine, has two distinct meanings. The fundamental balance of an engine is related to the basic position and number of the cylinders. Thus a six cylinder in-line engine can be designed to be in perfect balance. A four cylinder in-line engine always has a secondary out-of-balance force (i.e. at twice engine frequency). A four cylinder opposed piston 4-cycle engine always has an unbalanced rocking couple. These are all inherent design features of the particular configuration and, as such, cannot be changed by the tuner. Balance weights are sometimes forged or cast integrally with the crankshaft. These are to counterbalance some of the rotating mass and to reduce the peak loads on the main bearings. This again is a balancing feature that is designed into the engine and is normally beyond the scope of the tuner.

The balancing that comes into our operating field is the process by which we match the rotating masses of crankshaft, connecting rods and pistons to help smooth out the variations in centrifugal and inertia that exist between cylinders on a production engine. One sometimes hears that the act of balancing an engine makes an incredible improvement over the standard engine. This is seldom true, but the perfectionist, who regards a 2 per cent improvement as a major triumph, will be well satisfied. A well-tuned engine, with carefully matched combustion chamber volumes and with piston rings and bores in excellent condition, will inevitably have well balanced compression pressures. Such an engine deserves to have the inertia loads balanced too.

For the pistons and connecting rods all that is required is a scale balance with a sensitivity of $\frac{1}{4}$ gram. With such a balance available it is a simple matter to file or grind small amounts of metal from the inside face of the bottom edge of the piston skirt or from the bottom side of the gudgeon pin boss until all piston weights match the weight of the lightest within a total tolerance of $\frac{1}{2}$ gram. Connecting rods should be balanced to the same total tolerance, after crack detection and polishing. Metal should never be removed from the flanges of the connecting rod (i.e. the 'down-strokes' of the H section). The best zone for the removal of surplus metal is generally at the big-end. When new rods are to be fitted permission should be obtained from the store-keeper at the local parts stockist to borrow at least a dozen rods. From these a set of rods can be chosen for retention that are much closer to the same weight than is usually obtained by random selection. In this way the amount of metal to be removed becomes small and the danger of weakening the rod at a vital spot disappears.

Crankshafts should be crack-tested, polished over the whole surface of each web, then balanced. The flywheel should be bolted in place and the crankshaft, flywheel and clutch pressure-plate balanced as an assembly. Static balancing on knife-edges is better than nothing, but the serious competitor is well advised to send the crankshaft and flywheel assembly to be dynamically balanced by a firm of specialists. When a crankshaft is rotating the disposition of the out-of-balance masses relative to the centre-line of the shaft becomes of major importance. This factor is neglected when a crankshaft is balanced by simply rocking backwards and forwards on knife-edges.

WORK ON THE VALVE GEAR

Push rods

Anything we can do to lighten the valve gear without weakening it will all serve to reduce valve gear stresses. Push rods are a good example. Some makers offer a competition push rod incorporating a light-alloy tube in place of the standard steel tube or rod. These are always a good buy. Standard push rods should be checked for straightness. Any push-rod that is bowed by more than $0.010''$ measured at the centre should be replaced by a new one.

Tappets

On standard engines these components are often far too heavy. A little thought will soon show where metal can be removed with safety.

Valve rockers

The conventional rocker is of H-form in cross section. Metal can be judiciously ground from the web portion, i.e. the vertical faces, but the top and bottom flanges should be left intact, apart from the removal of excrescences and a light polish all over. In general one cannot recommend the drilling of lightening holes along the web since this can sometimes introduce stress concentrations ('stress-raisers') that will eventually lead to failure. One additional treatment to increase fatigue life is shot-peening. This is well worthwhile on a competition engine. It should precede final polishing.

Valve guide clearances

Since a hard-driven engine runs with higher temperatures throughout it is customary to increase bearing clearances, piston clearances, almost all clearances in components that operate at above-normal temperatures on a competition engine. A slight increase in valve guide clearance is permissible, but an excessive clearance can lead to trouble. More inlet valves stick through too much clearance than too little. A large clearance leaves a space for the build-up of deposits on the valve stem. In the case of the inlet valve these deposits take the form of a sticky gum produced by the breakdown of unsaturated hydrocarbons in the fuel. On the exhaust valves the deposits are largely composed of lead compounds from the T.E.L. in the fuel. These are not as sticky as the gum that forms on the inlet stems and are not as liable to cause sticking. Without previous knowledge, then, of the high-output operation of the valve train it is safest to leave the inlet valve guide clearances at standard and to increase the clearance in the exhaust valve guides by no more than 0·001".

Valve end clearances

Normal maker's recommended valve clearances are not usually sufficient in the case of the exhaust valve, when an engine is to be driven in competition. An increase of about 0·002" over the standard clearance is advisable; even more on certain air-cooled engines. When a change has been made to an austenitic material for the exhaust valve, as much as 0·004" increase in clearance is sometimes advisable, since this material has a greater coefficient of expansion than normal valve steels. Always check on this point when the valve specification is changed.

Increased valve clearance for better acceleration

A useful tip for sprint events or short duration races on twisty courses is to open up the valve clearances by about 0·005″. The loss of such a small amount in total valve lift has a negligible effect on the power output, but the valve opening duration is shortened to the benefit of low and medium speed torque. The effect is most noticeable when a 'full-race' type of camshaft has been fitted. With a conservative timing as on the Three Litre Austin-Healey the effect is negligible at low and medium speeds and actually loses a little power at the peak. It is a tip to remember, but the effect should be a matter for experiment during race practice.

Valve springs

Never take a chance on a valve spring. Failure here can mean a dropped valve, a broken piston or even a rod through the crankcase. Dual and triple springs give increased protection and the different natural frequencies of the two or three springs that are compounded can help to prevent valve bounce at high engine speeds. Competition engine springs should be shot-peened to increase fatigue life. A spring that has lost its initial spring set and is noticeably shorter than standard should be replaced by a new spring. If the tuner has a valve spring tension tester, each spring should have the load checked when compressed to fitted length. This should not vary by more than ±5 per cent over the set. When fitted to the cylinder head, each valve spring should be checked in turn, with the appropriate cam at full-lift, to see that a 0·002″ feeler can be passed between the coils. If any spring becomes coil-bound at full-lift a very high shock load can be thrown on the valve train. A buckled push-rod could be the result. With an experimental camshaft or any other untried modifications to the valve gear, the engine should be turned over very slowly when checking for coil-bound springs, since damage could be done even at this stage.

CONNECTING RODS

Connecting rods should always be checked for parallelism and twist before engine assembly. If the little-end bush has been honed in a modern machine, such as the Van Norman 'No. 232 Pin Shop', the gudgeon pin will sit in perfect alignment with the crank-pin.

When the little-end bush has been reamed or honed by less reliable methods we can check alignment by means of V-blocks or by the use of one of the special jigs sold to the motor trade. Acceptable limits for alignment are shown in Fig. 125. Special tools for straightening connecting rods are also sold to the motor trade. The writer

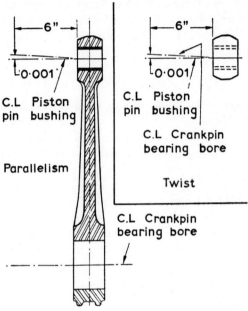

Fig. 125.   The acceptable limits of parallelism and twist

does not recommend their use on competition engines. A poorly aligned rod, provided it passes a crack-detection test and has no previous history of maltreatment (such as might be caused by a dropped valve or seized piston), should have a new bush fitted and then honed to size on a machine that ensures good alignment.

A final check on alignment should be made after assembly of the rods and pistons in the bores, with the crankshaft fitted and the big-ends fitted, torqued down and split-pinned. The engine should be turned over slowly while a mechanic looks up the bores to observe the behaviour of the little-end, between the inner faces of the gudgeon pin bosses on the piston. Each cylinder must be checked in turn to see that adequate clearance exists through a complete engine revolution between the little-end and its piston bosses.

Connecting rod bolts

New big-end bolts should always be fitted when an engine is re-built. One can hardly afford to gamble on the fatigue life of such a vital component. Little-end clamp bolts tend to stretch with repeated use. On a competition engine it pays to change these too at every engine strip.

PISTONS

On most sports cars to-day die-cast aluminium pistons are fitted as standard. For hard competition work the strength of the cast alloy is sometimes inadequate at the higher operating temperatures and pressures. A change to a forged material becomes necessary. The advice of the piston manufacturers should be sought in all cases. Solid-skirt pistons are stronger than split-skirt. For racing engines the closer fit and control of piston-slap given by the split-skirt piston must be sacrificed in the interests of reliability.

Many of the ills suffered by pistons are caused by defects in other components, or from mistakes during assembly. Here are a few useful tips.

Misalignment

When a piston has seized or shows signs of scuffing to make high pressure marks above the gudgeon pin hole on one side and below the gudgeon pin hole on the other side, the fault is almost sure to lie in an out-of-line boring of the little-end bush, not in the piston.

The gudgeon pins and corresponding holes in the piston are finished to very close tolerances, usually to a total tolerance of no more than 0·0002″. Pins are selectively fitted to pistons to achieve this and great care must be taken when assembling an engine to see that pins and pistons do not get interchanged. Pistons should be heated in hot oil or near-boiling water until the pins slide in easily by hand. Tapping with a hammer or drift might open the end of the pin slightly, leaving a high spot that could lead to trouble.

Petrol washing

Excessive and prolonged use of the choke can wash the lubricant off the cylinder walls. In mild cases the piston skirt will be slightly marked and the surface blackened with a mixture of carbon and iron dust from the rings. In bad cases a general seizure will occur all over the rings, lands and skirt. This again is an example of piston trouble not caused primarily by any deficiency in the piston or rings.

Circlips

When gudgeon pins are located by means of a circlip check that the circlip has fully expanded into its groove. A hung-up circlip can overstress the edge of the groove in the piston with subsequent failure of the metal at this point.

Ring gaps

Ring gaps on standard engines are never less than 0.003″ per inch of bore on water-cooled engines and 0·004″ per inch of bore on air-cooled engines. The maker's recommended gap should be increased by 0·001″ per inch of bore on rally cars and 0·002″ per inch of bore on racing cars. Negligible loss of compression occurs from a large ring gap, but the damage incurred when the ring ends butt can be catastrophic. This will be no ordinary seizure. Portions of the ring-belt usually break-up in the process.

SPARKING PLUGS

Heat range

It is customary to refer to a plug as hot or cold to indicate its capacity for conducting heat away from the central electrode through the insulator and the steel body of the plug to the metal of the cylinder head. The hot plug is the one with the greater resistance to the flow of this heat. This degree of resistance to heat flow is built into the plug design, usually in the manner shown in Fig. 18 of Chapter Three. The path from the insulator tip to its seat in the plug body is greater for the hot plug than the cold one. Actually the critical temperature zone, the surface of the insulator in the region of the electrode tip, should be in the same temperature range for all plugs, ideally in the narrow range 700-750°C.

All plug makers supply plugs in a wide range of heats. Occasionally a maker fails to cover one part of the range adequately. The tuner then refers to a comparison chart to find the maker who does supply the correct plug (see Appendix 2). The standard test for plug type suitability is to drive the car on full-throttle—or at least full-throttle for about three-quarters of the time—for several minutes. Two laps of a racing circuit would suffice. A clean cut of power, straight from full-throttle, is essential if the true evidence is to be left on the plug insulators. The accelerator should be lifted, the ignition cut and the car put into neutral, all inside two seconds. If the plug heat is correct for the conditions the colour of the insulator will be light brown. If it is off-white with a speckly appearance it has been operating at too high a temperature and should be replaced by a plug that is one or two steps colder in the heat range. If it is chalky white in colour it has been running much too hot and will require a change to a plug several steps colder. A dry sooty appearance, however, indicates that the plug insulator is running too cool. To burn off this film of carbon will require a plug that is one or two steps hotter. If the soot is oily in character it is more than likely that excessive oil consumption is the cause.

Long reach plugs are used on many British sports cars and these seem to have a marked tendency to pre-ignite under hard-driving conditions. Occasionally 'running-on' or 'dieseling' occurs after the ignition has been switched off. A change to a colder plug can be the answer, but this can increase the tendency to plug-fouling in traffic. Before committing oneself to a cooler plug it is advisable to examine the threads on all the plug bodies and at the base of the plug sockets, Running-on can be caused by overheating of a ragged edge on the screw thread. Some plug makers produce alternative plugs of identical reach and heat-range, the only difference being in the length of the threaded portion of the plug body. The omission of the bottom two turns of thread can sometimes prevent running-on. Engines with a known tendency to running-on should have the two bottom threads of the plug sockets filed down and polished when the cylinder head is off.

Plug gaps

The correct plug gap is often a matter for compromise. If the gap is too large misfiring can occur at high speed; if too small difficult starting can result. One can only experiment until the best results are given. In general, the effect of modification work is to make the demands on the ignition system more severe. Higher compression ratios and better breathing means that the density of the gas between the plug electrodes at the time of ignition is much higher. A reduction in gap size is sometimes necessary to keep the plug sparking at the high speed end of the range. The more one thinks about it, the more one sees a real need in the modern high-output engine for a new ignition system. Necessity must indeed be the Mother of Invention since novel ignition systems are announced almost every day!

Plug gaskets

The copper and asbestos gasket supplied with the majority of British sparking plugs is excellent from the gas sealing point of view, but does not rate too highly as a vehicle for the transfer of heat from the plug body to the cylinder head. Solid copper gaskets should always be used on an engine used for racing. Much greater attention to cleanliness of the plug socket sealing face is required, but the improved heat conductivity given by copper gaskets sometimes makes the difference between a smooth engine and a pre-igniting engine.

Cleaning

There is nothing wrong with garage plug cleaners if they are used intelligently. An oily plug should be cleaned first in petrol and then dried by air blast before being subjected to the sand-blast. After sand-blasting great care should be taken to see that all sand has been blasted from the narrow space at the top of the insulator. A little probing with a stiff fine wire will soon show if any particles are still lodged in this crevice.

Plug tightening torques

Before fitting, the plug threads should be smeared lightly with an anti-seize compound containing colloidal graphite or molybdenum disulphide.

The following tightening torques are recommended:

TABLE 9

(All torques expressed in lb.-ft.)

For cast-iron heads			
10 m.m. plugs		14 m.m. plugs	
fitted cold	fitted hot	fitted cold	fitted hot
14	14	30	30
For aluminium heads			
10 m.m. plugs		14 m.m. plugs	
fitted cold	fitted hot	fitted cold	fitted hot
11	7	27	23

EXHAUST VALVE MATERIALS

When the car manufacturer supplies larger valves as an item of stage tuning one can usually assume that the quality of the material has been suitably improved from standard. An increase in exhaust valve size inevitably means an increase in running temperature. Most of the heat passes out of the valve head through the seat. An increase in head diameter increases the surface to receive heat by a greater amount than the increase in seat length. Moreover an increase in inlet and exhaust valve diameters gives an increase in volumetric efficiency with a corresponding increase in the mass of hot gas to be exhausted through the valve.

The exhaust valve leads a very hard life, blasted on its head by burning gases at temperatures of about 2000°C during the firing stroke, surrounded by gases at about 1000°C during the exhaust stroke and only momentarily cooled on one side of the head by a draught of cool mixture during the induction stroke. Measurements made by Lyn and Spiers (*Trans. I.A.E.*, July 1947) on a typical water-cooled petrol engine showed the maximum temperature of the valve head to be about 750°C at maximum power output. Even at half-throttle the temperature remained above 600°C. At these temperatures mild steel would be plastic and would rapidly be drawn into the valve throat under the pull of the valve spring.

The writer has seen a 'tulip' valve produced in this way when unsuitable material was used.

XB type steels have been popular for a long time for medium powered production engines. This alloy contains about 20 per cent of chromium with small amounts of nickel, silicon and manganese. Higher performance engines call for a change to an austenitic steel such as En 54, which has roughly equal amounts of chromium and nickel at around 14 per cent, 2-4 per cent of tungsten and smaller amounts of silicon and manganese. For the racing engine Nimonic 80 is typical of the modern high-nickel alloy in which iron is almost completely displaced from the mixture. The analysis of this famous gas turbine alloy is given in Table 11 at the end of the chapter.

At room temperature the XB steel is nearly twice as hard as Nimonic 80. With this alloy and most austenitic steels too it is necessary to provide a special hard tip to the valve stem to prevent excessive wear when cold. At 550°C XB steel has become softer than Nimonic 80 and at 750°C it is as soft as a white metal bearing alloy. Table 10 below shows how Nimonic 80 maintains its hardiness at elevated temperatures, an essential quality in an exhaust valve material if the seat is not to be easily damaged.

TABLE 10

Material	Brinell hardness at			
	15°C	550°C	650°C	750°C
XB steel	400	180	80	40
En 54	169	138	104	*
Nimonic 80	233	200	183	164

* Value not available

Exhaust valve failures

One can group the types of failure experienced by exhaust valves under five headings.

(a) *Stem failure*. The hottest part of an exhaust valve is not usually the centre of the outer face of the head. Bedale and Graham (*Proc. I, Mech. E.*, 1956) found that for the valve used in their experiments the junction of the parallel stem and the beginning of the

SCE U

head radius was the hottest part of the valve and was 20°C hotter than the centre of the head face. It is thus at the top of the valve stem, i.e. the part close to the head radius, where one must expect failure to take place. To aggravate matters this region is prone to corrosion fatigue. Examination of exhaust valves during overhaul sometimes reveals a slight necking in this region. While it is an obvious step to scrap such a valve, it is not as obvious that this necking can be caused by corrosion and not necessarily by the plastic flow of the material near its yield point. Austenitic steels of the type used for exhaust valves can suffer an intergranular attack (corrosion seeping along the boundaries of the crystals) when in the presence of certain acids and when *at the same time* subjected to stress. This phenomenon is called 'stress corrosion'. The junction of the head and stem is subjected to fairly high alternating stresses and is at the same time in contact with hot, dilute, hydrochloric and hydrobromic acids. Young (*Eaton-Forum*, December issue, 1954) has shown that the hot stem is more prone to attack from another chemical, the lead oxide formed from the combustion of tetra-ethyl lead in 'leaded' fuels. Even Nimonic 80 is not immune to attack from stress corrosion. In all cases there appears to be a critical temperature at which lead oxide attack commences. For En 54 this temperature is approximately 725°C and for Nimonic 75 and 80 about 850°C, which fortunately is above the usually operating temperature. Chronic cases of stem breakage in an engine using austenitic steel exhaust valves are usually cured by a change to the more expensive Nimonic 80.

(*b*) *Stem-stretching and head deformation.* Insufficient strength at the operating temperature can result in stem stretching or 'tuliping' of the head. Both types of deformation are caused by flowing of the hot material under the inertia loads imposed as the valve springs pull it back on to its seat. A reduction in the closing velocity is beneficial, but a change to a material with an improved hot-strength is a more positive cure.

Stretching or tuliping in a valve can usually be spotted when a valve loses its clearance after only a short running time. Failure to restore the correct clearance will of course prevent the valve seating, which will soon lead to burning of the valve.

(*c*) *Cracking of the head.* The commonest form of crack found in exhaust valves follows a radial path, from the seat towards the centre. The radial path is produced by the stress pattern in the head, which is circumferential in character. As soon as a small crack appears in the valve seat, the hoop-stresses in the head, assisted by the blow-torch effect of the flames which escape through the crack, gradually open the crack wider and wider until severe burning and loss of power render that particular cylinder inoperative. The initial crack can be started in several ways. Attack of the valve material by lead oxide will produce a brittle scale. If a piece of this scale breaks away from the valve seat, a bad contact is then made with the seat on the head over the section of the seat affected. This permits a slight gas leakage which starts the notch that leads to eventual major failure. Another way is for combustion products to be trapped under the seat and to prevent full circumferential contact. The use of a narrow seat on the cylinder head is one way to combat this danger. A narrow seat exerts a higher seating pressure and tends to flatten and disperse any trapped carbon particles. Since the major heat-path from the valve to the cylinder head is through this seat the use of a very narrow seat will offer a greater resistance to heat flow than one of normal width. The choice of seat width is thus a compromise between conflicting requirements.

(d) *Guttering of the valve seat.* By guttering we mean that severe oxidation or melting of the metal over an arc of the seat that makes the valve seat appear to have been gnawed by a metal-eating rodent. This sad condition can again be the result of lead oxide scale fracture, or again be caused by the trapping of hard carbon particles. Sometimes it is caused by worn valve guides producing irregular and imperfect seat contact. Sometimes it is caused by distortion of the seat in the cylinder head. This is a design fault and is difficult to cure at a later stage. Inadequate provision of cooling water around the valve seat can sometimes lead to steam pockets forming in the water jacket on one side of the valve seat. Any collecting of steam bubbles in this way will seriously reduce the rate of heat transfer at this point. Increased water circulation at this point is the only cure.

(e) *Valve-sticking.* This is usually caused by the build-up of deposits on the valve stem. The critical part of the stem is that part which is exposed to the exhaust gases while the valve is open, but which enters the guide as the valve closes. Several years ago it

TABLE 11 EXHAUST VALVE MATERIALS

Make and Equivalent B.S. Spec.	Duty	Advantages and Disadvantages	Composition %		Coefficient of Thermal Expansion (20-600°C) ×10⁶
En 53	Production touring engine	Good creep resistance up to 700°C. Scaling resistance poor at 750°C.	C 0·55-0·65 Mn 0·30-0·60 Cr 5·75-6·75	Si 1·40-1·70 Ni 0-0·50 S & P 0-0·5	13·6
En 59 XB Steel	Acceptable for Fast touring	Good resistance to lead corrosion. Can be hardened.	C 0·74-0·84 Mn 0·20-0·60 Cr 19·00-20·50	Si 1·75-2·25 Ni 1·15-1·65 S & P 0-0·03	12·3
En 54 Jessop G2A D.T.D. 49B	High duty. Suitable for high output sports car	Good resistance to corrosion and lead attack. Good strength and hardness at high temp. Needs hard face to stem tip.	C 0·42 Mn 0·70 Cr 13·0 Nb 0·2	Si 1·30 Ni 13·0 W 2·50	18·9
En 55 Jessop R18 D.T.D. 282	As above	Slightly better resistance to lead attack and scaling than En 54.	C 0·35 Ni 8·0 W 3·50	Si 1·80 Cr 18·0	17·5
Jessop G32	For racing sports car engines	Very high creep and fatigue strength at 750-800°C. Expensive.	C 0·30 Ni 12·5 V 2·8	Si 0·5 Cr 19·0 Nb 1·3 Mn 0·8 Co 45·0 Mo 2·0	16·4
Nimonic 80 D.T.D. 725	Racing engines	Very high resistance to lead attack and corrosion. High strength at high temp. Expensive.	C 0·0-1 Cr 18-21 Fe 0·5-0 Si 1·0 max. Ti 1·8-2·7	Co 0·2-0 Mn 1·0 max. Al 0·5-1·8 Ni remainder	14·0

was thought to be good design to shield this portion of the stem by extending the guide into the port, this portion of the guide being counterbored, thus acting as a shield to the stem. The medicine was worse than the illness. The deposits gradually filled up the counterbored space and, since the exposed portion of the guide tended to run hotter than a flush guide, the deposits hardened quickly and increased the tendency to sticking.

We are not going to attempt to list all available exhaust valve materials, but Table 11 will serve to illustrate the ascending scale of available materials, giving their compositions and their limitations.

APPENDIX 1
Special Camshafts

THE SPECIAL grinds listed below are obtainable from the following suppliers or may be ordered from local speed shops.

CUSTOM CAM GRINDERS

B.M.C. — Competition Dept.
1200 Van Ness Avenue
San Francisco, California

Hollywood Sports Cars
5766 Hollywood Blvd.
Hollywood 28, California

Howard's Cams, Dept. S
10122 S. Main Street
Los Angeles, California

Ed. G. Iskenderian
607 N. Inglewood Ave.
Inglewood, California

or from the British Agent:
Motor Books
41/42 Parliament St.
Whitehall, London, S.W.1.

Potvin: Moon Equipment Co.
10820 S. Norwalk Blvd.
Santa Fe Springs, California

Racer Brown Camshaft Engineering Co.
8687-G Melrose Avenue
Los Angeles 46, California

Rich Motors
1615 S. Brand Blvd.
Glendale 1, California

Weber Speed Equipment
310 S. Center, Dept. G
Santa Ana, California

ERRATUM

The Address of Motor Books is now
Motor Books and Accessories
33 St. Martin's Court, London, W.C.2

CAM GRINDS

Cam	Duration Degrees	Timing IO BTC	IC ABC	EO BBC	EC ATC	Lift-clearance	Comment
						AUSTIN-HEALEY BN-4, BN-6	
Factory stock	230	5	45	40	10	·356″ @ ·013″ in. ·014″ ex.	Mild. Good mid range torque, Rev. limit: 5200.
Factory option No. H.8339	240	10	50	45	15	·390″ @ ·015″	More h.p. throughout range, top end extended only slightly. 6-port head & compression increase advised.
Weber Full Race	260	23	57	57	23	·409″ @ ·015″	Street & competition. Rough idle. More at top end and extended range. Spring spacers required. Supertuning of engine advised.
Racer Brown 16-A	270	26	64	64	26	·430″ @ ·014″	Competition. Sacrifice low end for good mid range & top. 6500 r.p.m. attainable with HD springs and supertuned engine.
Racer Brown SS6	278	26	72	75	23	·417″ @ ·014″	Competition, street. Some low end sacrifice, but very strong from 3000 to 6500. HD6 S.U. Carburettors, full length exhaust pipes recommended.
Isky T-36 T-3-6	264	27	57	57	27	·430″ @ ·018″ in. ·020″ ex.	Competition & street. Better above 2500 than stock without kit. With kit top end is extended appreciably.
Isky T-4-6	274	32	62	62	32	·430″ @ ·018″ in. ·020″ ex.	Competition. 4000 r.p.m. & up only.
Potvin X 95	278	31	67	70	28	·411″ @ ·014″	Supertuned engine, kit necessary. Street & competition. Power increase throughout normal range.
Potvin Eliminator	284	31	73	73	31	·420″ @ ·014″	Competition. Improved engine only. Top end extended, low range impaired.
Wilson No. 704	268	24	64	64	24	·420″ @ ·015″	Competition. Supertuned engine required. Strong from 3000 r.p.m. up.

Cam	Duration Degrees	IO BTC	Timing IC ABC	EO BBC	EC ATC	Lift-clearance	Comment
AUSTIN-HEALEY 3-LITRE							
Factory stock	230	5	45	40	10	·314" @ ·012"	Low end torque, good, sharp power drop above 5200 r.p.m.
Factory option (Mk. II)	230 in. 252 ex.	5	45	51	21	·356" @ ·013" in. ·014" ex.	Top end better, low range good. Intermediate cam for street-competition.
Howard A-6-64	263	28	56	56	28	·432" @ ·018" in. ·020" ex.	Mild grind with good idle for street or track.
Howard A-6-3	274	26	66	66	26	·326" @ ·018" in. ·020" ex.	Competition only with fully modified engine.
Factory option No. AEC	252	16	56	51	21	·356" @ ·013"	Competition street. Power range moved up, extended. Supertuned engine necessary.
Racer Brown ST15	278	31	67	71	21	·440" @ ·017"	Competition, straightaway speed runs. Over 205 h.p. with supertuned engine but effective only above 5000.
HSC 'G'	274 in. 268 ex.					·405" @ ·014"	Peak output from 195 h.p. @ 5500 r.p.m. with supertuned engine. Competition. 4000 to 6000. Requires 3 HD6 S.U.s.
AUSTIN-HEALEY SPRITE (AND M.G. MIDGET)							
Factory stock incl. 1961	230	5	45	40	10	·280" @ ·019"	Good for economy. Low end satisfactory with supertuned engine.
Factory Mk. II	230 in. 252 ex.	5	45	51	21	·312" @ ·021"	Better top & increased power through full range with factory supertuned engine.
Factory option Part 2A. 948	252	16	56	51	21	·310" @ ·015"	Good street cam for fully supertuned engine. Street-competition. Idle rough, low end sacrifice. HD springs required.
Factory option Part Q 2629	280	20	80	50	50	·380" @ ·015"	Competition. Long course. Still drivable on street. Rough idle & poor low end torque. HD springs required.

Cam						Lift @ clearance	Remarks
Potvin X 95	278	31	67	70	28	.411" @ .014"	Competition-street. Moderate engine supertuning required. Kit not necessary.
Potvin X 105	280	30	70	70	30	.411" @ .015"	Competition. Improved engine mandatory. Strong from 3000 up. Kit not required.
Weber 'Full Race' CTS	263	25	58	58	25	.330" @ .019"	Competition-street. Idle tolerable, 6500 r.p.m. with good power. Supertuned engine required.
Howard A-11	252	17	55	55	17	.324" @ .014" in. / .016" ex.	Comes in at 2300 with slightly rough idle. Strong to 7200 r.p.m.
Howard A-11A	261	20	61	61	20	.330" @ .017" in. / .020" ex.	Competition cam with good mid-range and a top end to 7400 r.p.m.
Howard A-16B	264	22	62	62	22	.330" @ .017" in. / .020" ex.	3800-8000 r.p.m. range. Strictly a track cam used with modified engine.
Isky MM-3	252	16	56	56	16	.320" @ .012" in. / .014" ex.	Street-competition. Low end impaired but strong acceleration over 3000. Modified engine, kit advised. Springs required.
Isky MM-32	257 in. 260 ex.	18	59	60	20	.325" @ .016" in. / .019" ex.	Competition-street. Rough idle. Power comes on at 3500. Rev. limit 7500 with modified engine. Kit required.
Isky MM-55	260	20	60	60	20	.325" @ .016" in. / .019" ex.	Competition-Formula Junior. Range: 4000 to 8000. Modified engine, kit required.
Racer Brown 244-A	295	40	75	79	36	.370" @ .017"	Competition-Formula Junior. Fast course. Modified engine only. Kit advised.
Racer Brown 268-A	276	26	70	74	22	.415" @ .017"	Competition. High-lift for High-revs. Better mid range than 244-A. Modified engine and kit required.
B.M.C.-Winfield 1000 308	308	46	82	82	46	.360" @ .019"	Competition-Formula Junior. Top end good. Modified engine necessary.
B.M.C.-Winfield 1100 316	316	50	86	86	50	.366" @ .019"	Competition-Formula Junior. Comes on at 4500, peak 6800, limit 8200 in fully modified engine.

Cam	Duration Degrees	Timing				Lift-clearance	Comment
		IO BTC	IC ABC	EO BBC	EC ATC		
			CHEVROLET CORVETTE				
Factory 'Regular' 1957-61 '283'	250	12.5	57.5	54.5	15.5	.398" O Lash	Hydraulic lifter cam needed for low end torque.
Factory 'Special' 1957-61 '283' 1962 '327'	287 299	35 32	72 87	76 74	31 45	.394" @ .008" in. .400" @ .018" ex.	Street-competition. 'Duntov' grind used in all solid lifter engines. An extremely successful camshaft.
Isky E-4	260	20	60	58	22	.410" @ .015"	Street-competition. For carburettor engines. 10% increase in h.p. above 2500 r.p.m. in moderately improved engine peak: 5500 r.p.m.
Isky Z-20	270	25	65	65	25	.420" @ .018"	Competition-street for modified carburettor-type engine. Peak 6000 r.p.m. Kit required.
Isky Z-30	280	30	70	70	30	.420" @ .018"	Competition. Much modified engine. Peak 6500, limit 7000. Idles at 1000. Kit required.
Racer Brown Super Street No. 4	285	31	74	78	27	.412" @ .012"	Competition-street. Injected or carburettor-fed modified engine. Top excellent, low and mid-range fair. No kit.
Racer Brown Super Street No. 3	261	21	60	60	21	.430" @ .014"	Street-competition for modified carb. or injected engine. Wide range under torque curve. Rough idle. No kit.
Racer Brown Super Torque No. 14	266	25	61	65	21	.485" @ .012"	Competition. Injected, modified engines. Good for slow courses. 2500-6000 r.p.m., strong. Limit 6500 with kit.
Racer Brown Super Torque No. 15	278	31	67	71	21	.485" @ .012"	Competition. 327 & bigger engines. Top end improved. Fast courses. 3000-6500 r.p.m., strong. Max. 7500 with kit.
Wilson No. 704	276	28	68	68	28	.420" @ .015"	Competition. Modified carburettor or injected engine. Best from 3000 to 6500 r.p.m. Low end down. No kit.

Cam	Duration				Lift	Remarks
Wilson No. 706	286	33	73	73	33	.435" @ .017"	Competition. Fully modified engine. Power strong from 3500 up. 7500 plus, OK. No kit.
Weber 500	266	23	58	58	23	.290" @ cam	Competition-street. Good idle. Good power over 4200. Modified engine with compression increase recommended.
Weber Hydro-Cheater	287	20	67	67	20	.275" @ cam	Street, drags. 14% increase in h.p. @ 4800 stock. Modified engine helps.
Potvin 153-B	284	32	72	74	30	.405" @ .014"	Competition. Top end improved and extended. Low range does not suffer appreciably. Injected or modified engine only.

FERRARI 250 GT

Cam							Remarks
Berlinetta or California non-competition	275 in. / 272 ex.	26	69	73	19	.360"	Adequate for touring.
Competition cam for all 3 litre V12s	301 in. / 290 ex.	45	75	70	40	.400"	Power much improved, r.p.m. range increased with improved head.

FIAT-ABARTH

Cam							Remarks
Abarth 850 Record Monza, 750 GT	280	30	70	70	30	.250" @ .012"	Strong mid range, rough on valve train at top end. Modified head required.
Rich Motors RM-E	297	36	81	81	36	.268" @ .012"	Competition only. Modified engine, mod. head necessary.
Rich Motors RG-E2	285 in. / 281 ex.	37	68	72	33	.250" @ .012"	Competition. Excellent in 5000 to 7500 range, poor below 4800. Mod. head & engine required.

FORD 105-E, 109-E (LOTUS MK. 7 SUPER)

Cam							Remarks
Factory stock 105E	240 in. / 236 ex.	10	50	44	10	.289" in. / .290" ex.	Morgan 4/4 Series III uses this one.
Factory option 105E	285 in. / 276 ex.	28	77	68	28	.374"	Street-competition. Much more through-out, but modifications almost necessity.
Cosworth 109E	316	50	86	86	50	.390"	Competition. Used in Lotus 7 Super. Modified engine required.

Cam	Duration Degrees	Timing				Lift-clearance	Comment
		IO BTC	IC ABC	EO BBC	EC ATC		
Isky Super RPM	280	35	65	65	35	·400" @ ·020"	Competition-junior. Modified engine required. Peak output 7000-7500. Range 4500-8000 under curve.
Racer Brown 244-A	295	40	75	79	36	·370" @ ·017"	Competition-junior. Good fast-course camshaft. High power peak, extended range.
Racer Brown 268-A	276	26	70	74	22	·415" @ ·017"	Competition-junior. Better mid-range torque than 244, high still good. Mod. engine required.
JAGUAR							
Factory stock XK120	252	15	57	57	15	·312" @ ·006" in. ·008" ex.	Stock.
Factory C Type (140-150) Part C 5717 (in.) Part C 5718 (ex.)	252	15	57	57	15	·375" @ ·006" in. ·006" ex.	Improved area under torque curve. In conjunction with porting, moved peak up.
Factory D Type (XKE) Part C 8512 (in.) Part 8513 (ex.)	270	30	60	60	30	·375"	Higher revs, higher peak h.p. Improved head and larger carburetors required.
Racer Brown 348-A	266	22	64	64	22	·335" @ ·018"	Street. More torque through mid-range than C type for 3·4 or 3·8.
Racer Brown 402-A	269	21	68	68	21	·402" @ ·018"	Street-competition. Mid-range, top end improved, range extended. Stock springs can be used.
Racer Brown 433-A	252	15	57	57	15	·422" @ ·018"	Competition; can be driven on street. Same timing as stock but high lift. Modified engine advised.

Racer Brown 460-A	272	26	66	66	26	·450" @ ·018"	Competition only. Top end of power curve extended. Fully modified engine required.
Isky X5	254	17	57	57	17	·390" @ ·012" in. / ·014" ex.	Street-competition. More acceleration and top end power.
Isky XM-2	254	17	57	57	17	·404" @ ·012" in. / ·014" ex.	Competition. Rev. limit extended, power peak moved up 500 r.p.m. Stock springs suitable. Modified engine required.
Isky XM-3	264	22	62	62	22	·404" @ ·012" in. / ·0.4" ex.	Competition. Fully mod. engine required. Power peak up 900 r.p.m. Poor torque below 4500 r.p.m.
Weber F6 H-J	264	22	62	66	18	·390" @ ·008"	Competition-street. Goes well with XK140 and 150. Approx. 20% increase in power & 7000 r.p.m. attainable in modified engine.
MGA							
Factory stock 1500 & 1600 MK II	252	16	56	51	21	·357" @ ·0·7"	Peak output at 5750 r.p.m., valve crash at 6000.
Factory stock 1600 Optional 1500 (1H 603)	230	5	45	40	10	·350" @ ·0·7"	Better low end & mid-range power. Some loss between 5000 & 6000
Factory option Part AEH 714	266	24	64	59	29	·350" @ ·017"	Better mid and top end. Modified engine advised. HD springs required.
Potvin X95	278	31	67	70	28	·411" @ ·014"	Street-competition. Increased power from 2000 r.p.m. upwards. Improved engine not required for street. No kit.
Potvin X105	280	30	70	70	30	·411" @ ·014"	Competition. Radical top end. Modified engine required. No kit.
Potvin Eliminator	284	31	73	73	31	·420" @ ·013"	Competition only with fully modified engine. Top end extended.
Weber CT M4	263	25	58	58	25	·420" @ ·019"	Competition-street. 15% gain in h.p. at peak. Rev. limit 6500 with modified engine. Stronger springs required.
Isky T-3	257	18	59	59	18	·428" @ ·014" in. / ·019" ex.	Street. Increase noted above 2500, peak at 6000. Modified engine not mandatory, but kit recommended.

Cam	Duration Degrees	IO BTC	Timing IC ABC	EO BBC	EC ATC	Lift-clearance	Comment
Isky T-32	257 in. 260 ex.	18	59	60	20	.428" in. .420" ex.	Competition. Short track, slow corners. 3500-7000 best range. Modified engine. Kit recommended.
Isky T-55	260	20	60	60	20	.420" @ .018" in. .019" ex.	Competition, longer courses, fast turns. Best range: 3500-7500. Kit required.
Racer Brown 17-A	270	28	62	66	24	.428" @ .017"	Competition, some street. Sacrifice low end for strong mid & top. Good at 6000 plus. Modified engine, stronger valve springs required.
Wilson 704	276	28	68	68	28	.420" @ .017"	Competition. Some street. All above 3000. Modified engine necessary.
Wilson 706	286	33	73	73	33	.435" @ .017"	Competition. Fully modified engine gives good torque above 3500 r.p.m.; poor below.
MG TD TF: 1250-1500							
Stock TC TD	247 in. 256 ex.	10	57	52	24	.320" @ .019"	Stock.
1250 TF Stock 1500 TF Stock Factory option Part AEG 122	230	5	45	45	5	.332" @ .012"	More low range torque than TD.
Weber 'Road Race'	264	22	62	66	18	.357" @ .014"	Competition-street. Peak at 6000, limit: 6500 with HD springs, modified engine.
Isky OA-1	241	12	49	48	13	.385" @ .015"	Competition-street. Best range 3500-6500 with modified engine. Kit required.
Howard M-13	248	15	53	54	16	.390" @ .016"	For maximum power from unmodified engine in the 2800-5700 r.p.m. range.
Howard M-13A	243	13	50	51	14	.390" @ .016"	Slightly rough idle but extends power range to 6100 for competition.

MORGAN PLUS 4

Cam					Lift	Remarks
Factory TR3 stock	250	15	55	55	·375" @ ·010"	Good low and mid-range.
Option 1961 model	299	43	76	76	·405" @ ·010"	Excellent improvement throughout range above 2500 when used with modified engine.
Potvin Eliminator	284	31	73	73	·420" @ ·014"	Competition. Fully modified engine only. Top end gain.
Potvin X95	278	31	67	70	·411" @ ·014"	Competition-street. Raises revs to 7000, some low end torque sacrifice. Modified engine recommended.

PORSCHE (Pushrod)

Cam					Lift	Remarks
Howard J-190	267	18	59	59	·389" @ ·011" in. / ·013" ex.	Improved cam primarily for street use but works well with modified engine.
Howard J-210	258	16	58	58	·402" @ ·013" in. / ·015" ex.	High lift, good torque for competition grind.
356 N-1500	220	2·5	37·5	37·5	·320"	Excellent performance. Improved engine necessary to take advantage of it. 1600 Models with roller cranks use S-1500 253° cam.
356 S-1500	253	19	54	54	·360"	
356/A & B N-1600	228	5	43	43	·334"	
356/A & B S-1600	245	15	50	50	·378" ex. / ·364" in.	
Racer Brown 22-A	254	19	55	58	·360" ex. / ·400" in. Super 90	Competition-street. Good slow course cam. 2500-6000 range. Max: 6800. Special springs required.
Racer Brown 21-A	253	14	59	60	·360" ex. / ·400" in. Super 90	Competition-street. Top end power shift. 3000-6500 range best. 6200 limit if stock springs are used.
Racer Brown 23-A	263 in. / 260 ex.	21	62	62	·365" ex. / ·407" in. Super 90	Competition only. Broad working range. 7000 plus in fully modified engine.
Isky 107	285	35	70	70	·365" ex. / ·375" in.	Street-competition. Broad range of power, peak at 5500. Special springs required.

Cam	Duration Degrees	Timing IO BTC	IC ABC	EO BBC	EC ATC	Lift-clearance	Comment
Isky 108	295	40	75	75	40	·400"	Competition, but street use possible.
Weber CF-P4	275 in. 274 ex.	28	67	74	20	·305" or ·340" optional	Competition-street. Modified engine necessary. 7000 r.p.m. limit.
Howard P-16	287	36	71	71	36	·363" @ ·007" in. ·008" ex.	Mid-range performance for both street and competition.
Howard P-16A	297	41	76	76	41	·402" @ ·007" in. ·009" ex.	A wild one for racing only with modified engine.
SUNBEAM ALPINE							
Factory stock	246	14	52	56	10	·366" in. ·012 ·364" ex. ·014	Smooth mild, good idle. Rev. limit 6000.
Factory option Part 10208620	264	25	59	63	21	·366" in. ·364" ex.	Relatively mild but carries power curve up to 7000 range. 8000 limit with certain engine mods.
Isky SB-2	264	22	62	62	22	·420" @ ·018"	Street-competition. Power moved up to 3500-6000 range. Some loss of low end power with modified engine. Limit raised to 7500 r.p.m. but stronger valve springs required.
TRIUMPH TR 3-4							
Factory stock TR2, 3, 3A & 4	250	15	55	55	15	·375" @ ·010"	Competition-street. Power increase through range, top end better, extended with modified engine. Stronger springs recommended.
Factory option TR4	278	31	67	70	28	·411" @ ·010"	

Name							Notes
Racer Brown 16-A	270	25	64	64	26	.430" @ .017"	Competition-street. High-lift takes this beyond street use. Low end drops off. Idle rough. Stronger springs lift r.p.m. to 6500.
Weber CT-T4	263	25	58	58	25	.420" @ .017"	Competition. Good top end, 6500 with shimmed springs. Modified engine recommended.
Isky TR 234	256	23	64	64	23	.425" @ .018"	Competition. Poor torque below 4000 r.p.m. Modified engine recommended. Kit not required.
Isky TR 555	270	25	65	65	25	.425" @ .018"	Competition. Peak around 5000, good to 6500. Modified engine, kit required.
Potvin X95	278	31	67	70	28	.411" @ .014"	Competition-street. Increase through full range, extend top end. With modified engine r.p.m. increased to 7000. (Not same as Option TR4.)
Howard T-193	251	16	55	55	16	.425" @ .016" in. / .017" ex.	Street use with stock engine. Good mid-range.
Howard T-193A	268	23	65	65	23	.425" @ .018"	Mild competition with power in 3000-6000 r.p.m. range.
Howard T-193B	272	26	66	66	26	.426" @ .019"	All-out for competition with good acceleration. Range 3300-6500 r.p.m.
VOLVO							
Factory stock P1800	268	24	64	62	26	.350" @ .018"	Street-competition. Better from 2500 up. 7000 available with modified engine. Kit recommended.
Isky vv-51	264	22	62	62	22	.400" @ .018"	
Isky vv 61	274	27	67	67	27	.400" @ .018"	Competition. Fully modified engine best for this top-end cam. Power good above 4500 r.p.m. Kit mandatory.
Potvin Eliminator	284	31	73	73	31	.420" @ .014"	Competition. Terrific top end. Fully modified engine necessary.

309

Champion Engineering Racing Division Heat Range Comparison Chart

Thread and Reach	Heat Range	Champion Regular	Champion Projected Core	Champion Racing	AC	Autolite	Lodge	KLG	Bosch
14 m.m. ⅜″ reach	HOT → COLD	J-6, J-6J	J-9Y	J-67J, J-64J	M44C, 43, 42S	AT-3, A3X, A-3, AR-32, A-32	2 HAN-14, HAN14P, HANP, HAN	F-250 (with spacer)	W225T3
		J-4, J-4J		J-63T, J-63R	M-42, C42M, M-42G, 42MS, M42K	AT-2, AT22, A23, A703, A2X, A21X			
				J-61R, J-59T, J-58T, J-58R, J-56T, J-55T	M-41, C42-1, M-41G, 41MS	A-901, A-201, A-603	R-47 (with spacer)	F-260 (with spacer)	W240T1
				J-53T	C-42	A-403	R-49 (with spacer)	F-280 (with spacer)	
						A-203	R-51 (with spacer)	F-320 (with spacer)	
14 m.m. ¾″ reach	HOT → COLD	N-5		N-63R	R44XL, 43M, C44XL Com., C42N	AG4, AGR41, AG42, AGR-42, AG2, AG3, AG22, AG23, AGR31, AGR32, AG901, AG201	HLNP, HLN, HLNY	FE50, PFE70, PFE50, FE100	W225T2, W240T17, W260T20, W280T2, W310T17, W340T17, W400T17
		N-3		N-58R, N-55T, N-53T	AG603, AG701, AG403, AG203		RL-47, RL-49, RL-51	FE-280, FE-290, FE-320	

Spark plug heat-range cross-reference chart. Columns are manufacturer code series (no brand labels printed); rows are grouped by seat/reach type, arranged HOT → COLD.

Reach / Seat	Heat	(1)	(2)	(3)	(4)	(5)	(6)	(7)	(8)
Taper seat	HOT	F-10, F-82	F-83Y	F-63R	C83T	BTF-3, BF32, BF22	HTN-18, HTN18P		W225T1
Taper seat	COLD		F-62Y	F-58R, F-55T, F-53T		BTF-1, BF601, BF703, BF603, BF403, BF203			
14 m.m. ½″ reach	HOT	L-7, L-85*, L-81*		L-63R	44F, 44FF	AE4, AE42	HH-14	CL-5, F-80	W270T16
14 m.m. ½″ reach	↓	L-5			43F	AE2, AE22, AE23, AE703, AE22	2HN-14	F-220	W240T16
14 m.m. ½″ reach	↓			L-59T, L-58R, L-56T, L-55T, L-53T	42FF, 42L Com.	AE603	R-47	F-250	W310T16
14 m.m. ½″ reach	COLD					AE403, AE203	R-49, R-51	F-280, F-310	W370T16, W400T16
18 m.m. ½″ reach	HOT	D-9, D-6		K-61R, K-59T, K-58R, K-55R, K-53T	C82, 81S Com., C81, M83, 83M	BT3, BT2, B903	1847, 1849, 1851	M250-4, M280-4, M290-4	M175T1, M225T1, DM270T16
18 m.m. ½″ reach	COLD				M82, 82M, M81, 81M	B603, B403, B203			DM310T1
18 m.m. ⅝″ reach	HOT			K-83R†, K-80R†		BN603, BN403			
18 m.m. ⅝″ reach	COLD			K-78T†		BN203			

* .472 reach

† Available only through Engineering Racing Division.

APPENDIX 3

Suggested Maximum Torque Values

For Bright Cap Screws
(Mild Steel) (Cold Rolled)
Approx. 50% of Ultimate Strength

Bolt Nominal Diam. (in.)	Threads per inch	For 30,000 psi Bolt Stress	
		Torque (ft.-lb.)	Compression (lb.)
$\frac{1}{4}$	20	4	810
$\frac{5}{16}$	18	8	1350
$\frac{3}{8}$	16	12	2040
$\frac{7}{16}$	14	20	2790
$\frac{1}{2}$	13	30	3780
$\frac{9}{16}$	12	45	4860

For Dark or Black Heat Treated Screws
Approx. 38% of Ultimate Strength
For ASTM Spec. A-325 Material

Stud Nominal Diam.	Threads per inch	For 45,000 psi Bolt Stress	
		Torque (ft.-lb.)	Compression (lb.)
$\frac{1}{4}$	20	6	1215
$\frac{5}{16}$	18	12	2025
$\frac{3}{8}$	16	18	3060
$\frac{7}{16}$	14	30	4185
$\frac{1}{2}$	13	45	5670
$\frac{9}{16}$	12	68	7290

Size			
5/8	11	60	6060
3/4	10	100	9060
7/8	9	160	12570
1	8	245	16530
1 1/8	7	390	20760
1 1/4	7	545	26700
1 3/8	6	730	31620
1 1/2	6	875	38820
1 5/8	5 1/2	1200	45450
1 3/4	5	1550	52320
1 7/8	5	2100	61470
2	4 1/2	2250	69000

Size			
5/8	11	90	9090
3/4	10	150	13590
7/8	9	240	18855
1	8	368	24795
1 1/8	8	533	32760
1 1/4	8	750	41805
1 3/8	8	1020	51975
1 1/2	8	1200	63225
1 5/8	8	1650	75600
1 3/4	8	2250	89100
1 7/8	8	3000	103680
2	8	3300	119340

Bolts must be well lubricated. Values are based on bolts lubricated with a heavy oil and graphite mixture.

APPENDIX 4

S.A.E. Horse Power

Standard conditions

Barometric pressure	29·92 in. Hg
Vapour pressure	Zero (dry air)
Carb. air temperature	60°F

Conditions of test are as follows in general

1. No air cleaner.
2. No generator.
3. No engine fan (unless engine is air-cooled).
4. No exhaust heat to intake manifold.
5. Exhaust system is dual piping from exhaust manifold, leading to an evacuated laboratory exhaust system giving a back pressure not less than atmospheric or more than 1 inch Hg.
6. The ignition timing and the mixture strength are adjusted at each engine speed (every 200 r.p.m.) to give the optimum power.
7. The standard water pump and fuel pump are fitted and the power to drive them is therefore included in the S.A.E. standard h.p.

The base engine power and torque as measured on the dynamometer, with the above concessions as to removal of equipment, is then corrected to standard S.A.E. conditions to give the standard S.A.E. figures.

Power correction factor

$$\text{Corrected b.h.p.} = [(\text{observed b.h.p.} + \text{friction h.p.}) \times \text{correction factor}] - \text{friction horse power}$$

$$\text{Correction factor} = \frac{29 \cdot 92}{B - E} \times \sqrt{\frac{460 + t}{520}}$$

where B = Observed barometric pressure, inches **Hg**.
E = Vapour pressure, inches **Hg**.
t = Intake air temperature, °F.

314

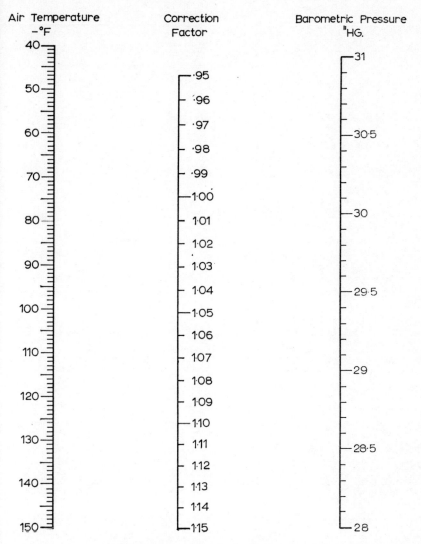

Air Temperature –°F	Correction Factor	Barometric Pressure "HG.
40		31
50	·95	
	·96	
60	·97	30·5
	·98	
70	·99	
	1·00	
80	1·01	30
	1·02	
90	1·03	
	1·04	29·5
100	1·05	
	1·06	
110	1·07	
	1·08	29
120	1·09	
	1·10	
130	1·11	28·5
	1·12	
140	1·13	
	1·14	
150	1·15	28

S.A.E. correction factor for horse power

Index